# The Analysis of Cross-Classified Categorical Data

# The Analysis of Cross-Classified Categorical Data

**Stephen E. Fienberg**

The MIT Press
Cambridge, Massachusetts, and London, England

Second printing, 1978


This book was set in Monophoto Times Mathematics and printed and bound in the United States of America.

Library of Congress Cataloging in Publication Data

Fienberg, Stephen E
The analysis of cross-classified categorical data.

Bibliography: p.
Includes index.
1. Multivariate analysis. 2. Contingency tables.
I. Title.
QA278.F53 519.5′3 76-58457
ISBN 0-262-06063-9

To Fred

# Contents

# Preface

The analysis of cross-classified categorical data has occupied a prominent place in introductory and intermediate-level statistical methods courses for many years, but with a few exceptions the only techniques presented in such courses have been those associated with the analysis of two-dimensional contingency tables and the calculation of chi-square statistics. During the past 15 years, advances in statistical theory and the ready availability of high-speed computers have led to major advances in the analysis of multidimensional cross-classified categorical data. Bishop, Fienberg, and Holland [1975], Cox [1970a], Haberman [1974a], Lindsey [1973], and Plackett [1974] have all presented detailed expositions of these new techniques, but these books are not directed primarily to the nonstatistical reader, whose background may be limited to one or two semesters of statistical methods at a noncalculus level.

The present monograph is intended as an introduction to the recent work on the analysis of cross-classified categorical data using loglinear models. I have written primarily for nonstatisticians, and Appendix I contains a summary of theoretical statistical terminology for such readers. Most of the material should be accessible to those who are familiar with the analysis of two-dimensional contingency tables, regression analysis, and analysis-of-variance models. The monograph also includes a variety of new methods based on loglinear models that have entered the statistical literature subsequent to the preparation of my book with Yvonne Bishop and Paul Holland. In particular, Chapter 4 contains a discussion of contingency tables with ordered categories for one or more of the variables, and Chapter 8 presents several new applications of the methods associated with incomplete contingency tables (i.e., tables with structural zeros).

Versions of material in this monograph were prepared in the form of notes to accompany lectures delivered in July 1972 at the Advanced Institute on Statistical Ecology held at Pennsylvania State University and during 1973 through 1975 at a series of Training Sessions on the Multivariate Analysis of Qualitative Data held at the University of Chicago. Various participants at these lectures have provided me with comments and suggestions that have found their way into the presentation here. Most of the final version of the monograph was completed while I was on sabbatical leave from the University of Minnesota and under partial support from National Science Foundation Grant SOC72-05257 to the Department of Statistics, Harvard University, and grants from the Robert Wood Johnson Foundation and the Commonwealth Fund to the Center for the Analysis of Health Practices, Harvard School of Public Health.

I am grateful to Stephen S. Brier, Michael L. Brown, Ron Christensen,

David R. Cox, William Fairley, S. Keith Lee, William Mason, and Roy E. Welsch for extremely valuable comments and suggestions. Many people have provided me with examples and other materials, from both published and unpublished works, that have found their way into the final manuscript, including Albert Beaton, Richard Campbell, O. Dudley Duncan, Leo Goodman, Shelby Haberman, David Hoaglin, Kinley Larntz, Marc Nerlove, S. James Press, Ira Reiss, Thomas Schoener, and Sanford Weisberg. Most of all, I am indebted to Yvonne Bishop, Paul Holland, and Frederick Mosteller, whose collaboration over a period of many years helped to stimulate the present work.

For the typing and organization of the final manuscript, I wish to thank Sue Hangge, Pat Haswell, Susan Kaufman, and Laurie Pearlman.

New Brighton, Minnesota
August 1976

Stephen E. Fienberg

# The Analysis of Cross-Classified Categorical Data

# 1
# Introduction

## 1.1 The Analysis of Categorical Data

A variety of biological and social science data come in the form of cross-classified tables of counts, commonly referred to as contingency tables. The units of a sampled population in such circumstances are cross-classified according to each of several categorical variables or sets of categories such as sex (male, female), age (young, middle-aged, old), or species. Intermediate-level statistics textbooks for biologists, such as Bliss [1967], Snedecor and Cochran [1967], and Sokal and Rohlf [1969], focus on the analysis of such data in the special case of two-way cross-classifications, as do textbooks for social scientists, such as Blalock [1972]. More detailed treatments, by Maxwell [1961] and by Fleiss [1973], are also available. A review of the material presented in one or more of these books is adequate preparation for this presentation.

When we look at several categorical variables simultaneously, we say that they form a multidimensional contingency table, with each variable corresponding to one dimension of the table. Such tables present special problems of analysis and interpretation, and these problems have occupied a prominent place in statistical journals since the first article on testing in $2 \times 2 \times 2$ tables by Bartlett [1935].

Until recent years the statistical and computational techniques available for the analysis of cross-classified data were quite limited, and most researchers handled multidimensional cross-classifications by analyzing various two-dimensional marginal totals, that is, by examining the categorical variables two at a time. This practice has been encouraged by the wide availability of computer program packages that automatically produce chi-square statistics for all two-dimensional marginal totals of multidimensional tables. Although such an approach often gives great insight about the relationship among variables, it

(a) confuses the marginal relationship between a pair of categorical variables with the relationship when other variables are present,
(b) does not allow for the simultaneous examination of these pairwise relationships,
(c) ignores the possibility of three-factor and higher-order interactions among the variables.

My intention here is to present some of the recent work on the statistical analysis of cross-classified data using loglinear models, especially in the multidimensional situation. The models and methods that will be considered do not have the shortcomings mentioned above. All the techniques described will be illustrated by actual data. Readers interested in mathematical proofs should turn to the source articles or books cited.

I view this monograph as an introduction to a particular approach to the analysis of cross-classified categorical data. For more details on this approach, including mathematical proofs, various generalizations, and their ramifications, see Bishop, Fienberg, and Holland [1975] or Haberman [1974a]. Other presentations with differing contents or points of view include Cox [1970a], Goodman [1970, 1971b], Grizzle, Starmer, and Koch [1969], Ku, Varner, and Kullback [1971], Lancaster [1969], Lindsey [1973], and Plackett [1974]. Bock [1970, 1975] also discusses the analysis of cross-classified data, based on the notion of multinomial response relationships much like those considered here.

## 1.2 Forms of Multivariate Analysis

The analysis of cross-classified categorical data falls within the broader framework of multivariate analysis. A distinction will be made here between variables that are free to vary in response to controlled conditions—that is, *response* variables—and variables that are regarded as fixed, either as in experimentation or because the context of the data suggests they play a determining or causal role in the situation under study—that is, *explanatory* variables. Dempster [1971] notes that the distinction between response and explanatory variables need not be firm in a given situation, and in keeping with this view, in exploratory analyses, we often choose different sets of response variables for the same data set.

Of importance in describing various types of models and methods for multivariate analysis is the class of values assumed by the variables being examined. In many circumstances, we wish to distinguish among variables whose values are

(i) dichotomous (e.g., yes or no),
(ii) nonordered polytomous (e.g., five different detergents),
(iii) ordered polytomous (e.g., old, middle-aged, young),
(iv) integer-valued (e.g., nonnegative counts), or
(v) continuous (at least as an adequate approximation).

Variables with values of types (i) through (iv) are usually labeled *discrete*, although integer-valued variables might also be treated as if they were continuous. Here the term categorical will be used to refer primarily to types (i), (ii), and (iii), and the possibility of type (iv) will be ignored. Mixtures of categorical and continuous variables appear in many examples.

We can categorize classes of multivariate problems by the types of response and explanatory variables involved, as in the cross-classification of Table 1-1.

**Table 1-1**
Classes of Statistical Problems

| | | Explanatory Variables | | |
|---|---|---|---|---|
| | | Categorical | Continuous | Mixed |
| Response Variables | Categorical | (a) | (b) | (c) |
| | Continuous | (d) | (e) | (f) |
| | Mixed | ? | ? | ? |

The cells in the bottom row of this table all contain question marks in order to indicate the lack of generally accepted classes of multivariate models and methods designed to deal with situations involving mixtures of continuous and discrete response variables. Dempster [1973] has proposed a class of logit models that is of use here, but his approach has yet to see much application. The cells in the middle row correspond to problems dealt with by standard multivariate analysis, involving techniques such as
(d) analysis of variance,
(e) regression analysis,
(f) analysis of covariance (or regression analysis with some dummy variables).

The work on linear logistic response models by Walker and Duncan [1967], Cox [1970a], and Haberman [1974a] deals with problems for all three cells in the first row when there is a single dichotomous response variable, while the more recent results of Nerlove and Press [1973] handle multiple response variables. Linear logistic response models will be discussed to some extent in Chapter 6. Cell (a) of the table corresponds to cross-classified categorical data problems, and some of the most widely used models for their analysis will be described in the following chapters.

The models used throughout this book rely upon a particular approach to the definition of interaction between or among variables in multidimensional contingency tables, based on cross-product ratios of expected cell values. As a result, the models are linear in the logarithms of the expected value scale; hence the label *loglinear* models. There are several analogies between interaction in these loglinear models and the notion of interaction in analysis-of-variance (ANOVA) models. These will be pointed out in the course of the discussion. The use of ANOVA-like notation is deceptive, however. In ANOVA models one tries to assess the *effects* of independent variables on a dependent variable and to partition overall variability. In contingency table analysis the ANOVA-like models are used to describe the structural relationship among the variables corresponding to the dimensions of the table. The distinction here is important, and the fact that many researchers have not understood it has led to considerable confusion.

When a distinction is made between explanatory variables and response variables, loglinear models can be converted into *logit* or *linear logistic* response models, in which one predicts log-odds quantities involving the dependent (or response) variables using a linear combination of effects due to the explanatory variables. There is a much closer analogy between these linear logistic models and the usual ANOVA or regression models. This point is discussed in considerable detail in Chapter 6.

## 1.3 Some Historical Background

The use of cross-classifications to summarize counted data clearly predates the early attempts of distinguished investigators such as Quetelet in the mid-nineteenth century to summarize the association between variables in a $2 \times 2$ table. Not until the turn of the century, however, did Pearson and Yule formulate the first major developments in the analysis of categorical data. Despite his proposal of the well-known chi-square test for independence for two-dimensional cross-classifications (see Pearson [1900a]), Karl Pearson preferred to view a cross-classification involving two or more polytomies as arising from a partition of a set of multivariate data, with an underlying continuum for each polytomy and a multivariate normal distribution for the "original" data. This view led Pearson [1900b] to develop his tetrachoric correlation coefficient for $2 \times 2$ tables and served as the basis for the approach adopted by many subsequent authors, such as Lancaster [1957] and Lancaster and Hamdan [1964]. This approach in some sense also led to Lancaster's method of partitioning chi-square, to which I shall return shortly. The most serious problems with Pearson's approach were (1) the complicated infinite series linking the tetrachoric correlation coefficient with the frequencies in a $2 \times 2$ table and (2) his insistence that it *always* made sense to assume an underlying continuum for a dichotomy or polytomy, even when the dichotomy of interest was dead–alive or employed–unemployed, and that it was reasonable to assume that the probability distribution over such a dead–alive continuum was normal.

Yule [1900], on the other hand, chose to view the categories of a cross-classification as fixed, and he set out to consider the structural relationship between or among the discrete variables represented by the cross-classification. This approach led him to consider various functions of the cross-product ratio, discussed here in Chapter 2. When there actually is an underlying continuum for each of two polytomies, the cross-product ratio for a $2 \times 2$

table resulting from a partitioning of the two variables simply is not a substitute for an estimate of the true correlation coefficient of the underlying continuum (see Plackett [1965] or Mosteller [1968]). Thus the methods proposed by Yule are not necessarily applicable in cases where the $2 \times 2$ table is simply a convenient summary device for continuous bivariate data and the original observations are in fact available.

The debate between Pearson and Yule was both lengthy and acrimonious (see, e.g., Pearson and Heron [1913]), and in some ways it has yet to be completely resolved, although the statistical literature of the past 25 years on this topic would indicate that Yule's position now dominates. In fact, Yule can be thought of as the founder of the loglinear model school of contingency table analysis, and most of the results in this book are an outgrowth of his pioneering work. However, the notions of Yule were not immediately generalized beyond the structure of two-dimensional tables. Thirty-five years passed before Bartlett [1935], as a result of a personal communication from R. A. Fisher, utilized Yule's cross-product ratio to define the concept of second-order interaction in a $2 \times 2 \times 2$ contingency table (see Chapter 3).

While the multivariate generalizations of the Yule–Bartlett cross-product ratio or loglinear model approach were fermenting, the technique of standardization (see Bishop, Fienberg, and Holland [1975], Chapter 4, and Bunker et al. [1969]) to eliminate the effects of categorical covariates received considerable attention in the epidemiological literature. Standardization is basically a descriptive technique that has been made obsolete, for most of the purposes to which it has traditionally been put, by the ready availability of computer programs for loglinear model analysis of multidimensional contingency tables. Thus it is not discussed in detail in this book.

During the past 25 years, the statistical literature on the analysis of categorical data has focused primarily on three classes of parametric models: (1) loglinear models, (2) additive models, and (3) models resulting from partitioning chi-square, which may be viewed as a combination of multiplicative and additive. This last class of models, which is usually associated with the work of Lancaster, is much misunderstood, as Darroch [1974, 1976] has recently noted. In addition, there has been a related literature on measures of association (e.g., Goodman and Kruskal [1954, 1959, 1963, 1972]). Although different groups of authors use different methods of estimation (maximum likelihood, minimum modified chi-square, or minimum discrimination information), almost all of the recent literature can be traced back either to the 1951 paper of Lancaster or to the work of S. N. Roy and his students at North Carolina in the mid-1950s (e.g., Roy and Kastenbaum [1956], Roy and Mitra [1956]). It is interesting that Roy's students developed his ideas using both the minimum modified chi-square approach (e.g., Bhapkar and Koch [1968], Grizzle, Starmer, and Koch [1969]) and the method of maximum likelihood (e.g., Bock [1970], Kastenbaum [1974]).

The major advances in the literature on multidimensional contingency tables in the 1960s grew out of Roy and Kastenbaum's work and papers by Birch [1963], Darroch [1962], Good [1963], and Goodman [1963, 1964]. These advances coincided with the emergence of interest in and the availability of high-speed computers, and this work received substantial impetus from several large-scale data analysis projects. Much of the recent literature on loglinear models can be linked directly to the National Halothane Study (see Bunker et al. [1969], Bishop, Fienberg, and Holland [1975], Mosteller [1968]), while problems in the Framingham Study led to work on linear logistic models involving both categorical and continuous predictor variables (e.g., Cornfield [1962], Truett, Cornfield, and Kannel [1967]). The Framingham study work paralleled work on linear logistic models by Cox [1966] and had ties to the earlier contributions of Berkson [1944, 1946].

A fairly complete bibliography for the statistical literature on contingency tables through 1974 is given by Killion and Zahn [1976].

## 1.4 A Medical Example

Table 1-2 presents data compiled by Cornfield [1962] from the Framingham longitudinal study of coronary heart disease (see Dawber, Kannel, and Lyell [1963] for a detailed description). Variable 1 is a binary response variable indicating the presence or absence of coronary heart disease, while variable 2

**Table 1-2**
Data from the Framingham Longitudinal Study of Coronary Heart Disease (Cornfield [1962])

| Coronary Heart Disease | Serum Cholesterol (mg/100 cc) | Systolic Blood Pressure (mm Hg) <127 | 127–146 | 147–166 | 167+ |
|---|---|---|---|---|---|
| Present | <200 | 2 | 3 | 3 | 4 |
| | 200–219 | 3 | 2 | 0 | 3 |
| | 220–259 | 8 | 11 | 6 | 6 |
| | ≥260 | 7 | 12 | 11 | 11 |
| Absent | <200 | 117 | 121 | 47 | 22 |
| | 200–219 | 85 | 98 | 43 | 20 |
| | 220–259 | 119 | 209 | 68 | 43 |
| | ≥260 | 67 | 99 | 46 | 33 |

(serum cholesterol at four levels) and variable 3 (blood pressure at four levels) are explanatory. The data as displayed in Table 1-2 form a $2 \times 4 \times 4$ three-dimensional contingency table.

This example is typical of those encountered in medical contexts. Although serum cholesterol and blood pressure might well be viewed as continuous variables, the values of these variables have been broken up into four categories each, corresponding to different levels of a priori perceived risk of coronary heart disease. An alternative to this approach would be to treat the variables as continuous and to use a regression-like logistic response model that expresses the dependency of coronary heart disease in a smooth and simple fashion.

# 2
# Two-Dimensional Tables

## 2.1 Two Binomials

We often wish to compare the relative frequency of occurrence of some characteristic for two groups. In a review of the evidence regarding the therapeutic value of ascorbic acid (vitamin C) for treating the common cold, Pauling [1971] describes a 1961 French study involving 279 skiers during two periods of 5–7 days. The study was double-blind with one group of 140 subject receiving a placebo while a second group of 139 received 1 gram of ascorbic acid per day. Of interest is the relative occurrence of colds for the two groups, and Table 2-1 contains Pauling's reconstruction of these data.

If $P_1$ is the probability of a member of the placebo group contracting a cold and $P_2$ is the corresponding probability for the ascorbic acid group, then we are interested in testing the hypothesis that $P_1 = P_2$. The observed numbers of colds in the two groups, $x_{11} = 31$ and $x_{21} = 17$ respectively, are observations on independent binomial variates with probabilities of success $P_1$ and $P_2$ and sample sizes $n_1 = 140$ and $n_2 = 139$. The difference in observed proportions,

$$\bar{P}_1 - \bar{P}_2 = \frac{x_{11}}{n_1} - \frac{x_{21}}{n_2},$$

has mean $P_1 - P_2$ and variance

$$\frac{P_1(1 - P_1)}{n_1} + \frac{P_2(1 - P_2)}{n_2}.$$

**Table 2-1**
Incidence of Common Colds in a Double-Blind Study Involving 279 French Skiers (Pauling [1971])

(a) Observed values

| | | Cold | No Cold | Totals |
|---|---|---|---|---|
| Treatment | Placebo | 31 | 109 | 140 |
| | Ascorbic Acid | 17 | 122 | 139 |
| | Totals | 48 | 231 | 279 |

(b) Expected values under independence

| | | Cold | No Cold | Totals |
|---|---|---|---|---|
| Treatment | Placebo | 24.1 | 115.9 | 140 |
| | Ascorbic Acid | 23.9 | 115.1 | 139 |
| | Totals | 48 | 231 | 279 |

If $P_1 = P_2$, then we could estimate the common value by

$$\bar{P} = \frac{\text{total no. of colds}}{n_1 + n_2} \tag{2.1}$$

and the estimated variance of $\bar{P}_1 - \bar{P}_2$ by

$$\bar{P}(1 - \bar{P})\left(\frac{1}{n_1} + \frac{1}{n_2}\right). \tag{2.2}$$

Assuming that the hypothesis $P_1 = P_2$ is correct, a reasonable test can be based on the approximate normality of the standardized deviate

$$z = \frac{\bar{P}_1 - \bar{P}_2}{\sqrt{\bar{P}(1 - \bar{P})\left(\frac{1}{n_1} + \frac{1}{n_2}\right)}}. \tag{2.3}$$

For our example we get

$$z = \frac{\frac{31}{140} - \frac{17}{139}}{\sqrt{\frac{48}{279} \times \frac{231}{279} \times \left(\frac{1}{140} + \frac{1}{139}\right)}} = 2.19,$$

a value that is significant at the 0.05 level. If we take these data at face value, then we would conclude that the proportion of colds in the vitamin C group is smaller than that in the placebo group. This study, however, has a variety of severe shortcomings (e.g., the method of allocation is not specified and the evaluation of symptoms was largely subjective). For a further discussion of these data, and for a general review of the studies examining the efficacy of vitamin C as a treatment for the common cold up to 1974, see Dykes and Meier [1975].

As an alternative to using the normal approximation to the two-sample binomial problem, we could use the Pearson chi-square statistic (see Pearson [1900a]),

$$X^2 = \sum \frac{(\text{Observed} - \text{Expected})^2}{\text{Expected}}, \tag{2.4}$$

where the summation is over all four cells in Table 2-1. We obtain the expected values by estimating $P_1 = P_2 = P$ (the null value) as $\bar{P} = 48/279$; that is, we multiply the two sample sizes $n_1$ and $n_2$ by $\bar{P}$, obtaining the expected values for the (1, 1) and (2, 1) cells, and then get the other two expected values by subtraction. Table 2-lb shows these expected values, and on substituting the observed and expected values in expression (2.4) we get $X^2 = 4.81$, a value that may be referred to a $\chi^2$ distribution with 1 d.f. (degree of freedom). A large value of $X^2$ corresponds to a value in the right-hand tail of the $\chi^2$ distribution and is indicative of a poor fit. Rather than using the $\chi^2$ table we note that the square root of 4.81 is 2.19, the value of our $z$-statistic computed earlier. Some elementary algebra shows that, in general, $z^2 = X^2$. If we set $x_{12} = n_1 - x_{11}$ and $x_{22} = n_2 - x_{21}$, then

$$z^2 = \frac{\left(\dfrac{x_{11}}{n_1} - \dfrac{x_{21}}{n_2}\right)^2}{\left(\dfrac{x_{11}+x_{21}}{n_1+n_2}\right)\left(\dfrac{x_{12}+x_{22}}{n_1+n_2}\right)\left(\dfrac{1}{n_1}+\dfrac{1}{n_2}\right)} \tag{2.5}$$

$$= \frac{[x_{11}(n_2 - x_{12}) - x_{21}(n_1 - x_{11})]^2(n_1+n_2)}{(x_{11}+x_{21})(x_{12}+x_{22})n_1 n_2}$$

and

$$X^2 = \frac{\left[x_{11} - n_1\left(\dfrac{x_{11}+x_{21}}{n_1+n_2}\right)\right]^2}{n_1\left(\dfrac{x_{11}+x_{21}}{n_1+n_2}\right)} + \frac{\left[x_{12} - n_1\left(\dfrac{x_{12}+x_{22}}{n_1+n_2}\right)\right]^2}{n_1\left(\dfrac{x_{12}+x_{22}}{n_1+n_2}\right)}$$

$$+ \frac{\left[x_{21} - n_2\left(\dfrac{x_{11}+x_{21}}{n_1+n_2}\right)\right]^2}{n_2\left(\dfrac{x_{11}+x_{21}}{n_1+n_2}\right)} + \frac{\left[x_{22} - n_2\left(\dfrac{x_{12}+x_{22}}{n_1+n_2}\right)\right]^2}{n_2\left(\dfrac{x_{12}+x_{22}}{n_1+n_2}\right)} \tag{2.6}$$

$$= \frac{[x_{11}(n_2 - x_{12}) - x_{21}(n_1 - x_{11})]^2(n_1+n_2)}{(x_{11}+x_{21})(x_{12}+x_{22})n_1 n_2}.$$

The use of the statistic $X^2$ is also appropriate for testing for independence in $2 \times 2$ tables, as noted in the next section.

Throughout this book we use the Greek quantity $\chi^2$ to refer to the chi-square family of probability distributions, and the Roman quantity $X^2$ to refer to the Pearson goodness-of-fit test statistic given in general by expression (2.4).

## 2.2 The Model of Independence

We have just examined a $2 \times 2$ table formed by considering the counts generated from two binomial variates. For this table the row totals were fixed

**Table 2-2**
Counts for Structural Habitat Categories for *sagrei* Adult Male *Anolis* Lizards of Bimini (Schoener [1968])

(a) Observed values

| | | Perch Diameter (inches) | | |
|---|---|---|---|---|
| | | $\leq 4.0$ | $> 4.0$ | Totals |
| Perch Height (feet) | $>4.75$ | 32 | 11 | 43 |
| | $\leq 4.75$ | 86 | 35 | 121 |
| | Totals | 118 | 46 | 164 |

(b) Expected values under independence

| | | Diameter | | |
|---|---|---|---|---|
| | | $\leq 4.0$ | $> 4.0$ | Totals |
| Height | $> 4.75$ | 30.9 | 12.1 | 43 |
| | $\leq 4.75$ | 87.1 | 33.9 | 121 |
| | Totals | 118 | 46 | 164 |

by design. In other circumstances we may wish to consider $2 \times 2$ tables where only the total for the table is fixed by design. For example, ecologists studying lizards are often interested in relationships among the variables that can be used to describe the lizard's habitat. Table 2-2a contains data on the habitat of *sagrei* adult male *Anolis* lizards of Bimini, originally reported by Schoener [1968] in a slightly different form. A total of 164 lizards were observed, and for each the perch height (variable 1) and perch diameter (variable 2) were recorded. The two variables were dichotomized partially for convenience and partially because the characteristics of interest for perches are high versus low and wide versus narrow.

Let us denote the observed count for the $(i, j)$ cell by $x_{ij}$ and the totals for the $i$th row and $j$th column by $x_{i+}$ and $x_{+j}$, respectively. Corresponding to this table of observed counts is a table of probabilities,

| | | Variable 2 | | |
|---|---|---|---|---|
| | | 1 | 2 | Totals |
| Variable 1 | 1 | $p_{11}$ | $p_{12}$ | $p_{1+}$ |
| | 2 | $p_{21}$ | $p_{22}$ | $p_{2+}$ |
| | Totals | $p_{+1}$ | $p_{+2}$ | 1 |

(2.7)

where the probabilities $\{p_{ij}\}$ add to 1, $p_{i+} = p_{i1} + p_{i2}$, and $p_{+j} = p_{1j} + p_{2j}$. In the two-binomial example we wanted to compare two proportions. In the present situation we wish to explore the relationship between the two dichotomous variables corresponding to rows and to columns, that is, to perch height and perch diameter.

If perch height is independent of perch diameter, then

$$\begin{aligned} p_{ij} &= \Pr\{\text{row category} = i \text{ and column category} = j\} \\ &= \Pr\{\text{row category} = i\} \Pr\{\text{column category} = j\} \\ &= p_{i+}p_{+j} \end{aligned} \tag{2.8}$$

for $i = 1, 2$ and $j = 1, 2$. Since only the total sample size $N$ is fixed, $\{x_{ij}\}$ is an observation from a multinomial distribution with sample size $N$ and cell probabilities $\{p_{ij}\}$. The expected value of $x_{ij}$ (viewed as a random variable) is $m_{ij} = Np_{ij}$, and under the model of independence $m_{ij} = Np_{i+}p_{+j}$. Finally, if we substitute the observed row proportion $x_{i+}/N$ for $p_{i+}$ and the observed column proportion $x_{+j}/N$ for $p_{+j}$, we get the well-known formula for the estimated expected value in the $(i, j)$ cell:

$$\hat{m}_{ij} = x_{i+}x_{+j}/N. \tag{2.9}$$

Table 2-2b displays the estimated expected values for the lizard data (assuming that perch height is independent of perch diameter). We can then test this hypothesis of independence using the Pearson chi-square statistic of expression (2.4):

$$\begin{aligned} X^2 &= \sum_{i=1}^{2} \sum_{j=1}^{2} \frac{(x_{ij} - x_{i+}x_{+j}/N)^2}{x_{i+}x_{+j}/N} \\ &= \frac{N(x_{11}x_{22} - x_{12}x_{21})^2}{x_{1+}x_{2+}x_{+1}x_{+2}}. \end{aligned} \tag{2.10}$$

For the data in Table 2-2 $X^2 = 0.18$, and comparing this value with a table for the $\chi^2$ distribution with 1 d.f., such as in Appendix III, we conclude that the model of independence of perch height and perch diameter fits the data quite well.

Table 2-3a presents a set of data similar to that in Table 2-2a, but for a different species of *Anolis* lizards, *distichus*, and a different size, adults and subadults. Were perch height and perch diameter independent here, we would expect the entries in the table to be as in Table 2-3b. The model of independence, if applied to the data in Table 2-3a, yields $X^2 = 1.83$, and again the model fits the data quite well.

**Table 2-3**
Counts for Structural Habitat Categories for *distichus* Adult and Subadult *Anolis* Lizards of Bimini (Schoener [1968])

(a) Observed values

| | | Perch Diameter (inches) | | |
|---|---|---|---|---|
| | | $\leq 4.0$ | $> 4.0$ | Totals |
| Perch Height (feet) | $> 4.75$ | 61 | 41 | 102 |
| | $\leq 4.75$ | 73 | 70 | 143 |
| | Totals | 134 | 111 | 245 |

(b) Expected values under independence

| | | Diameter | | |
|---|---|---|---|---|
| | | $\leq 4.0$ | $> 4.0$ | Totals |
| Height | $> 4.75$ | 55.8 | 46.2 | 102 |
| | $\leq 4.75$ | 78.2 | 64.8 | 143 |
| | Totals | 134 | 111 | 245 |

We note that $X^2$ as given by expression (2.10) is the same as $X^2$ in expression (2.6), provided we equate the totals in the corresponding tables, that is, provided we set

$$n_1 = x_{1+}, \quad n_2 = x_{2+}, \quad \text{and} \quad N = n_1 + n_2.$$

## 2.3 The Loglinear Model

In the two-dimensional tables just examined, the estimated expected values were of the form

$$\hat{m}_{ij} = \frac{x_{i+}x_{+j}}{N}, \qquad i = 1, 2, \quad j = 1, 2, \tag{2.11}$$

both under the model of independence of row and column variables and under the model of equality of binomial proportions (more generally referred to as the model of *homogeneity of proportions*). The hat in expression (2.11) is a reminder that $m_{ij}$ is a parameter being estimated by $\hat{m}_{ij}$.

Taking the natural logarithm of both sides of equation (2.11),

$$\log \hat{m}_{ij} = \log x_{i+} + \log x_{+j} - \log N, \tag{2.12}$$

and thinking in terms of an $I \times J$ table with $I$ rows and $J$ columns reveals a close similarity to analysis-of-variance notation. Indeed, the additive form suggests that the parameter $m_{ij}$ be expressed in the form

$$\log m_{ij} = u + u_{1(i)} + u_{2(j)}, \tag{2.13}$$

where $u$ is the grand mean of the logarithms of the expected counts,

$$u = \frac{1}{IJ}\sum_{i=1}^{I}\sum_{j=1}^{J} \log m_{ij}, \tag{2.14}$$

$u + u_{1(i)}$ is the mean of the logarithms of the expected counts in the $J$ cells at level $i$ of the first variable,

$$u + u_{1(i)} = \frac{1}{J}\sum_{j=1}^{J} \log m_{ij}, \tag{2.15}$$

and, similarly,

$$u + u_{2(j)} = \frac{1}{I}\sum_{i=1}^{I} \log m_{ij}. \tag{2.16}$$

Because $u_{1(i)}$ and $u_{2(j)}$ represent deviations from the grand mean $u$,

$$\sum_{i} u_{1(i)} = \sum_{j} u_{2(j)} = 0. \tag{2.17}$$

In the case of the model of homogeneity of proportions, $n_i = x_{i+}$ is fixed, and thus

$$\log \hat{m}_{ij} = -\log(n_1 + n_2) + \log n_i + \log x_{+j}. \tag{2.18}$$

The term $u_{1(i)}$ in the corresponding model of the form (2.13) is fixed by the sample design and is not otherwise directly interpretable.

It is rarely the case that the model of independence fits as well as it did in the two examples in the preceding section, and this result is all the more surprising in that the frequency of wide perches tends to decrease with increasing perch height in many natural habitats. If we think in terms of perch height and perch diameter interacting, we can add an "interaction term" to the independence model, yielding

$$\log m_{ij} = u + u_{1(i)} + u_{2(j)} + u_{12(ij)}, \tag{2.19}$$

where, in addition to (2.17), we have

$$\sum_{i=1}^{I} u_{12(ij)} = \sum_{j=1}^{J} u_{12(ij)} = 0. \tag{2.20}$$

For the example on vitamin C and the common cold, a term $u_{12(ij)}$ would represent the fact that the two binomial proportions are not equal. The model given by expressions (2.19), (2.17), and (2.20) is the most general one for the two-dimensional table.

## 2.4 Sampling Models

There are three commonly encountered sampling models that are used for the collection of counted cross-classified data:

(1) *Poisson:* We observe a set of Poisson processes, one for each cell in the cross-classification, over a fixed period of time, with no a priori knowledge regarding the total number of observations to be taken. Each process yields a count for the corresponding cell (see Feller [1968]). The use of this model for contingency tables was first suggested by Fisher [1950].

(2) *Multinomial:* We take a fixed sample of size $N$ and cross-classify each member of the sample according to its values for the underlying variables. This was the model assumed for the lizard examples of Section 2.3.

(3) *Product-Multinomial:* For each category of the row variable we take a (multinomial) sample of size $x_{i+}$ and classify each member of the sample according to its category for the column variable (the roles of rows and columns can be interchanged here). This was the sampling model used in the example on vitamin C and the common cold.

A more technical discussion of these sampling models will be given after multidimensional tables have been considered. The basic result of interest here is that the three sampling schemes lead to the same estimated expected cell values and the same goodness-of-fit statistics. (For a more general statement of this result, see the theoretical summary in Appendix II.)

Some stress has been placed on the sampling model in this section, but it is equally important to distinguish between response variables and explanatory variables. For the example of Table 2-1, which has a product-binomial sampling model, treatment (placebo or vitamin C) is an explanatory variable and the occurrence of colds is the response variable. For the two lizard examples, both perch height and perch diameter can be viewed as response variables, and the sampling model is approximately Poisson. It is also possible to have a Poisson or a multinomial sampling model with one response and one explanatory variable. For example, suppose we take a random sample of husband–wife pairs and cross-classify them by voting behavior (e.g., liberal or conservative) in the last election, with the husband's voting behavior as the row variable and the wife's voting behavior as the column variable. If we wish to assess the effect of the husband's behavior

**Table 2-4**
Piano Choice of Soloists Scheduled for the 1973–1974 Concert Season, for Selected Major American Orchestras

| Orchestra | Piano Choice Steinway | Other | Totals |
|---|---|---|---|
| Boston Symphony | 4 | 2 | 6 |
| Chicago | 13 | 1 | 14 |
| Cleveland | 11 | 2 | 13 |
| Minnesota | 2 | 2 | 4 |
| New York Philharmonic | 9 | 2 | 11 |
| Philadelphia | 6 | 0 | 6 |
| Totals | 45 | 9 | 54 |

on the wife, for example, the wife's behavior is a response variable and the husband's behavior an explanatory variable. If we condition on the husband's behavior, we go from a multinomial sampling model to a situation based on a product-multinomial model.

It must be noted that not all two-dimensional tables are necessarily generated by one of the three sampling models listed above. Table 2-4 presents data on the piano choice of soloists scheduled during the 1973–1974 concert season, for selected major American orchestras. For these data the basic sampling unit is the soloist; a given soloist may appear with several orchestras during a concert season, however, and in all appearances he or she will use the same brand of piano. Indeed, the total of 9 for "other" can probably be attributed to two or three pianists who use Baldwin pianos.

## 2.5 The Cross-Product Ratio and 2 × 2 Tables

In the examples above $I = J = 2$, and, because of the constraints,

$$u_{12(11)} = -u_{12(12)} = -u_{12(21)} = u_{12(22)}. \tag{2.21}$$

Thus there is, in effect, one parameter to measure interaction. When this parameter is set to zero, we get the one degree of freedom associated with the chi-square tests for independence or homogeneity of proportions. For $2 \times 2$ tables we can show that

$$u_{12(11)} = \tfrac{1}{4} \log \alpha, \tag{2.22}$$

where

$$\alpha = \frac{m_{11}m_{22}}{m_{12}m_{21}}. \tag{2.23}$$

We can also write

$$\alpha = \frac{p_{11}p_{22}}{p_{12}p_{21}}, \tag{2.24}$$

since the expected value for the $(i, j)$ cell is just the probability associated with that cell times the sample size:

$$m_{ij} = Np_{ij}. \tag{2.25}$$

This relationship between expected values and cell probabilities follows from results related to the sampling models considered above.

The quantity $\alpha$ defined by (2.23) or (2.24) is usually referred to as the *cross-product ratio* (see Mosteller [1968]) or *odds-ratio*, and it is a basic measure of association in $2 \times 2$ tables, in part because it is appropriate for all three types of sampling models described in Section 2.4. Surprisingly, this measure appears only rarely in social science research literature, although it is widely used in chemical, genetic, and medical contexts (see, e.g., Anderson and Davidovits [1975], Kimura [1965], and Cornfield [1956]). The cross-product ratio $\alpha$ has several desirable properties:

(1) It is invariant under the interchange of rows or columns (except for its "sign": if we interchange rows or columns but not both, the sign of $\log \alpha$ changes).

(2) It is invariant under row and column multiplications: Suppose we multiply the probabilities in row 1 by $r_1 > 0$, row 2 by $r_2 > 0$, column 1 by $c_1 > 0$, and column 2 by $c_2 > 0$, and then renormalize these values so that they once again add to 1. The normalizing constant cancels out and we get

$$\alpha' = \frac{(r_1c_1p_{11})(r_2c_2p_{22})}{(r_1c_2p_{12})(r_2c_1p_{21})} = \frac{p_{11}p_{22}}{p_{12}p_{21}} = \alpha. \tag{2.26}$$

(3) Clear interpretation: If we think of row totals as fixed, then $p_{11}/p_{12}$ is the odds of being in the first column given that one is in the first row, and $p_{21}/p_{22}$ is the corresponding odds for the second row. The relative odds for the two rows, or the odds-ratio, is then

$$\frac{p_{11}/p_{12}}{p_{21}/p_{22}} = \frac{p_{11}p_{22}}{p_{12}p_{21}} = \alpha. \tag{2.27}$$

(4) It can be used in $I \times J$ tables (and multidimensional tables) either through a series of $2 \times 2$ partitionings or by looking at several $2 \times 2$ subtables.

The quantity $\alpha$ runs from 0 to $\infty$ and is symmetric in the sense that two values of the cross-product ratio $\alpha_1$ and $\alpha_2$ such that $\log \alpha_1 = -\log \alpha_2$ represent the same degree of association, although in opposite directions.

If $\alpha = 1$, the variables corresponding to rows and columns are independent; if $\alpha \neq 1$, they are dependent or associated.

The observed cross-product ratio,

$$\hat{\alpha} = \frac{x_{11}x_{22}}{x_{12}x_{21}}, \tag{2.28}$$

is the maximum-likelihood estimate of $\alpha$ for all three types of sampling models considered above.

For the data in Tables 2-2a and 2-3a we estimate $\alpha$ as $\hat{\alpha}_1 = 1.18$ and $\hat{\alpha}_2 = 1.42$, respectively. The estimate of the large-sample standard deviation of $\log \hat{\alpha}$ is

$$s_\alpha = \sqrt{\frac{1}{x_{11}} + \frac{1}{x_{12}} + \frac{1}{x_{21}} + \frac{1}{x_{22}}}, \tag{2.29}$$

and we can use $\log \hat{\alpha}$ and $s_\alpha$ to get confidence intervals for $\log \alpha$, since for large samples $\log \hat{\alpha}$ is normally distributed with mean $\log \alpha$. Also, we can use the statistic

$$X^2 = \frac{(\log \hat{\alpha})^2}{s_\alpha^2} = \frac{(\log x_{11} + \log x_{22} - \log x_{12} - \log x_{21})^2}{\left(\frac{1}{x_{11}} + \frac{1}{x_{12}} + \frac{1}{x_{21}} + \frac{1}{x_{22}}\right)} \tag{2.30}$$

as an alternative to the usual chi-square statistic (2.10) to test for independence in a $2 \times 2$ table. Several authors have suggested alternative ways of getting confidence intervals and tests of significance for $\alpha$ or its logarithm. For an excellent review and discussion of these methods, see Gart [1971].

Goodman and Kruskal [1954, 1959, 1963, 1972] discuss various measures of association for two-dimensional tables, and in the $2 \times 2$ case a large number of these are simply monotone functions of the cross-product ratio. For example, Yule's [1900] $Q$ can be written as

$$\begin{aligned} \hat{Q} &= \frac{x_{11}x_{22} - x_{12}x_{21}}{x_{11}x_{22} + x_{12}x_{21}} \\ &= \frac{\hat{\alpha} - 1}{\hat{\alpha} + 1}. \end{aligned} \tag{2.31}$$

Pielou [1969] and others have been critical of such measures because they preclude the distinction between complete and absolute association. For example, Table 2-5 contains two cases in which $Q = 1$ (and $\alpha = \infty$). Pielou's claim is that any ecologist would assert that the association in Table 2-5b

**Table 2-5**
Examples of Complete and Absolute Association

(a) Complete association

| | | Species B Present | Absent | Totals |
|---|---|---|---|---|
| Species A | Present | 60 | 20 | 80 |
| | Absent | 0 | 20 | 20 |
| | Totals | 60 | 40 | 100 |

(b) Absolute association

| | | Species B Present | Absent | Totals |
|---|---|---|---|---|
| Species A | Present | 80 | 0 | 80 |
| | Absent | 0 | 20 | 20 |
| | Totals | 80 | 20 | 100 |

is greater by far than the association in Table 2-5a, and thus $Q$ (or $\alpha$) is not the most desirable measure in ecological contexts. Of course, measures that are not monotonic functions of $\alpha$ are tied into the relative values of the marginal totals, and this too is undesirable. For a further discussion of this point, see the papers by Goodman and Kruskal, or Bishop, Fienberg, and Holland [1975].

Returning to model (2.19) for the $2 \times 2$ table, it is of interest that the remaining two parameters, $u_{1(1)}$ $(= -u_{1(2)})$ and $u_{2(1)}$ $(= -u_{2(2)})$, are not simply functions of the row and column marginal totals, as one might have expected. In fact,

$$u_{1(1)} = \frac{1}{4} \log \frac{m_{11}m_{12}}{m_{21}m_{22}} \tag{2.32}$$

and

$$u_{2(1)} = \frac{1}{4} \log \frac{m_{11}m_{21}}{m_{12}m_{22}}. \tag{2.33}$$

All the subscripted parameters in the general loglinear model for two-dimensional tables can thus be expressed as functions of various cross-product-ratio-like terms. Nevertheless, $\alpha$ and the marginal totals $\{m_{i+}\}$ and $\{m_{+j}\}$ completely determine the $2 \times 2$ table.

## 2.6 Interrelated Two-Dimensional Tables

As we mentioned in Chapter 1, a typical analysis of a multidimensional cross-classification often consists of the analysis of two-dimensional marginal totals. Moreover, we noted that such analyses have severe shortcomings. To illustrate this point we now consider the analysis of a three-dimensional

**Table 2-6**
Two-Dimensional Marginal Totals of Three-Dimensional Cross-Classification of 4353 Individuals (See Table 3-6)

(a) Occupational group vs. educational level

| | (low) E1 | E2 | E3 | (high) E4 | Totals |
|---|---|---|---|---|---|
| O1 | 239 | 309 | 233 | 53 | 834 |
| O2 | 6 | 11 | 70 | 199 | 286 |
| O3 | 1 | 7 | 12 | 215 | 235 |
| O4 | 794 | 781 | 922 | 501 | 2998 |
| Totals | 1040 | 1108 | 1237 | 968 | 4353 |

O1 = self-employed, business
O2 = self-employed, professional
O3 = teacher
O4 = salaried, employed

(b) Aptitude vs. occupational level

| | O1 | O2 | O3 | O4 | Totals |
|---|---|---|---|---|---|
| (low) A1 | 122 | 30 | 20 | 472 | 644 |
| A2 | 226 | 51 | 66 | 704 | 1047 |
| A3 | 306 | 115 | 96 | 1072 | 1589 |
| A4 | 130 | 59 | 38 | 501 | 728 |
| (high) A5 | 50 | 31 | 15 | 249 | 345 |
| Totals | 834 | 286 | 235 | 2998 | 4353 |

(c) Aptitude vs. educational level

| | E1 | E2 | E3 | E4 | Totals |
|---|---|---|---|---|---|
| A1 | 215 | 208 | 138 | 83 | 644 |
| A2 | 281 | 285 | 284 | 197 | 1047 |
| A3 | 372 | 386 | 446 | 385 | 1589 |
| A4 | 128 | 176 | 238 | 186 | 728 |
| A5 | 44 | 53 | 131 | 117 | 345 |
| Totals | 1040 | 1108 | 1237 | 968 | 4353 |

example solely in terms of its two-dimensional marginal totals. Later we shall reconsider this example and contrast the conclusions drawn from the two analyses.

Table 2-6a contains data on the cross-classification of 4353 individuals into four occupational groups (O) and four educational levels (E). These data have been extracted from a larger study encompassing many more occupational groups. The Pearson chi-square test for independence of occupation and education yields $X^2 = 1254.1$ with 9 d.f., so that occupation and education are clearly related.

Tables 2-6b and 2-6c present two additional two-dimensional cross-classifications of the same 4353 individuals, the first by aptitude (as measured at an earlier date by a scholastic aptitude test) and occupation, the second by aptitude and education. Testing for independence in these tables yields, for Table 2-6b, $X^2 = 35.8$ with 12 d.f., and for Table 2-6c, $X^2 = 178.6$ also with 12 d.f. Both of these values are highly significant when referred to the corresponding chi-square distribution, and we are forced to conclude that occupation, aptitude, and education are pairwise related in all possible ways.

Now, in what ways is this analysis unsatisfactory? Everything we have done appears to be correct. We have just not done enough, and we have not looked at the data in the best possible way. Before we can complete the discussion of this example, however, we must consider details of the analysis of three-dimensional tables.

## 2.7 Correction for Continuity

The practice of comparing the Pearson chi-square statistic given by expression (2.4) to the tail values of the $\chi^2$ distribution with the appropriate degrees of freedom is an approximation that is appropriate only when the overall sample size $N$ is large. Yates [1934] suggested that a correction be applied to $X^2$, the *correction for continuity*, for $2 \times 2$ tables to make the tail areas correspond to those of the hypergeometric distribution (used when both row and column margins are fixed). Thus, instead of using the chi-square formula (2.10), many introductory texts suggest using the "corrected" statistic:

$$X_c^2 = \sum_{i=1}^{2}\sum_{j=1}^{2} \frac{[|x_{ij} - (x_{i+}x_{+j}/N)| - 1/2]^2}{x_{i+}x_{+j}/N}$$

$$= \frac{N(|x_{11}x_{22} - x_{21}x_{12}| - N/2)^2}{x_{1+}x_{2+}x_{+1}x_{+2}}. \tag{2.34}$$

Cox [1970b] provides an elementary analytical derivation of the continuity correction that is applicable in this case. Rao [1973, p. 414] provides a slightly different method for correcting $X^2$.

If, however, our aim is to correct the statistic $X^2$ so that it more closely adheres to the large-sample $\chi^2$ distribution, rather than to the hypergeometric distribution, then the use of the corrected chi-square statistic (2.34) may not necessarily be appropriate. In fact, Plackett [1964], Grizzle [1967], and Conover [1974] have shown that using $X_c^2$ in place of $X^2$ results in an overly conservative test, one that rejects the null hypothesis too rarely relative to the nominal level of significance. For example, in 500 randomly generated multinomial data sets examined by Grizzle, with $N = 40$ and cell probabilities

| | |
|---|---|
| 0.56 | 0.24 |
| 0.14 | 0.06 |

the statistic $X^2$ exceeded the 0.05 level of significance (i.e., 3.84) 26 times (5.2%) whereas the corrected statistic $X_c^2$ exceeded the 0.05 level about 6 times (1.2%).

Because of this empirical evidence, and because of our use of the $\chi^2$ distribution as the reference distribution for the chi-square statistic $X^2$, we make no use of continuity corrections in this book.

## 2.8 Other Scales for Analyzing Two-Dimensional Tables

The discussion in this monograph is focused on models for cell probabilities or expected values which are linear in the logarithmic scale. Other possible scales of interest are the linear scale and those based on the angular (arc sine) or integrated normal (probit) transforms. Another possibility is a model that is linear in the logistic scale; that is, if $p$ is a probability, its logistic transform is $\log(p/(1-p))$. In Chapter 6 the relationship between loglinear and linear logistic models is discussed in detail.

Cox [1970a, pp. 26–29] examines the linearizing transformations from $p(x)$, the probability of success at level $x$, to the following scales:

(1) logistic: $$\log\left(\frac{p(x)}{1-p(x)}\right), \tag{2.35}$$

(2) linear: $$p(x), \tag{2.36}$$

(3) integrated normal: $$\phi^{-1}(p(x)), \text{ where} \tag{2.37}$$

$$\phi(u) = \frac{1}{\sqrt{2\pi}} \int_{-\infty}^{u} e^{-y^2/2}\, dy;$$

(4) arc sine: $$\sin^{-1}(\sqrt{p(x)}). \tag{2.38}$$

Cox notes that all four transformations are in reasonable agreement when the probability of success is in the range 0.1–0.9, and that in the middle of this range analyses in terms of each of the four are likely to give virtually equivalent results.

For further details the interested reader is referred to Cox [1970a] and to the discussion in Bishop, Fienberg, and Holland [1975, pp. 368–369], which includes an example of a $2 \times 2 \times 2 \times 2$ table analyzed using both a loglinear or logistic model and an ANOVA model following an arc sine transformation.

# 3
# Three-Dimensional Tables

## 3.1 The General Loglinear Model

If we put Tables 2-2a and 2-3a together, we have a $2 \times 2 \times 2$ table with the dimensions perch height, perch diameter, and species (see Table 3-1). What are the interesting models for this $2 \times 2 \times 2$ table? To answer this we must start by introducing an appropriate notation.

Let $x_{ijk}$ be the observation in $i$th row, $j$th column, and $k$th layer of the table, and let $m_{ijk}$ be the corresponding expected value for that entry under some model. Our notation for marginal totals is such that when we add over a variable we replace the corresponding subscripts by a "+." Thus

$$x_{ij+} = x_{ij1} + x_{ij2}, \qquad i, j = 1, 2, \tag{3.1}$$

$$\begin{aligned} x_{i++} &= x_{i11} + x_{i12} + x_{i21} + x_{i22} \\ &= x_{i1+} + x_{i2+}, \qquad i = 1, 2, \end{aligned} \tag{3.2}$$

and

$$x_{+++} = \sum_{i=1}^{2} \sum_{j=1}^{2} \sum_{k=1}^{2} x_{ijk} = \sum_{i=1}^{2} x_{i++}. \tag{3.3}$$

Similarly, $m_{ij+} = m_{ij1} + m_{ij2}$, etc.

If the three variables corresponding to the dimensions of our table are independent, then

$$\begin{aligned} &\Pr\{\text{variable 1 takes level } i, \text{ variable 2 takes level } j, \text{ and variable 3 takes level } k\} \\ &= \Pr\{1 \text{ takes level } i\} \Pr\{2 \text{ takes level } j\} \Pr\{3 \text{ takes level } k\}, \end{aligned} \tag{3.4}$$

**Table 3-1**
Data from Tables 2-2a and 2-3a, along with Expected Values under Two Loglinear Models: Complete Independence (Column 3) and Conditional Indepdence of Variables 1 and 2 Given the Level of 3 (Column 4)

| (1) cell $(i, j, k)$ | (2) observed | (3) | (4) |
|---|---|---|---|
| (1, 1, 1) | 32 | 35.8 | 30.9 |
| (2, 1, 1) | 86 | 65.2 | 87.1 |
| (1, 2, 1) | 11 | 22.3 | 12.1 |
| (2, 2, 1) | 35 | 40.6 | 33.9 |
| (1, 1, 2) | 61 | 53.5 | 55.8 |
| (2, 1, 2) | 73 | 97.4 | 78.2 |
| (1, 2, 2) | 41 | 33.3 | 46.2 |
| (2, 2, 2) | 70 | 60.7 | 64.8 |

and by analogy with the model of independence in two dimensions, it would be natural for us to take our estimate of the expected count for the $(i, j, k)$ cell as

$$\hat{m}_{ijk} = \left(\frac{x_{i++}}{N}\right)\left(\frac{x_{+j+}}{N}\right)\left(\frac{x_{++k}}{N}\right)N. \tag{3.5}$$

Taking logarithms yields

$$\log \hat{m}_{ijk} = \log x_{i++} + \log x_{+j+} + \log x_{++k} - 2 \log N. \tag{3.6}$$

This additive form for the logarithms of the estimated expected values is once again highly reminiscent of analysis-of-variance notation. Switching from the estimated value $\hat{m}_{ijk}$ to the parameter $m_{ijk}$, if we set

$$u = \frac{1}{8}\sum_{i=1}^{2}\sum_{j=1}^{2}\sum_{k=1}^{2} \log m_{ijk} \tag{3.7}$$

$$= -2\log N + \frac{1}{2}\left[\sum_{i=1}^{2} \log m_{i++} + \sum_{j=1}^{2} \log m_{+j+} + \sum_{k=1}^{2} \log m_{++k}\right],$$

$$u_{1(i)} = \frac{1}{4}\sum_{j=1}^{2}\sum_{k=1}^{2} \log m_{ijk} - u, \tag{3.8}$$

etc., then we can write $\log m_{ijk}$ in an ANOVA-type notation:

$$\log m_{ijk} = u + u_{1(i)} + u_{2(j)} + u_{3(k)}, \tag{3.9}$$

where

$$\sum_{i=1}^{2} u_{1(i)} = \sum_{j=1}^{2} u_{2(j)} = \sum_{k=1}^{2} u_{3(k)} = 0, \tag{3.10}$$

since the terms involved represent deviations from the grand mean $u$.

Suppose now that the three variables are not independent. Four other types of models expressible in terms of the structural relationships among the three variables are as follows:

(1) Independence of (i) type of species and (ii) perch height and diameter jointly (there are two other versions of this model).
(2) Conditional independence of perch height and perch diameter given the species (again, there are two other versions of this model).
(3) Pairwise relations among the three underlying variables, with each two-variable interaction unaffected by the value of the third variable.
(4) Second-order interaction (or three-factor effect) relating all three variables,

so that the interaction between any two *does depend* on the value of the third variable.

All of the models listed above are special cases of what is known as the *general loglinear model:*

$$\log m_{ijk} = u + u_{1(i)} + u_{2(j)} + u_{3(k)} + u_{12(ij)} + u_{13(ik)} + u_{23(jk)} + u_{123(ijk)}, \tag{3.11}$$

where, as in the usual ANOVA model,

$$\begin{aligned} &\sum_i u_{1(i)} = \sum_j u_{2(j)} = \sum_k u_{3(k)} = 0, \\ &\sum_i u_{12(ij)} = \sum_j u_{12(ij)} = \sum_i u_{13(ik)} = \sum_k u_{13(ik)} \\ &\quad = \sum_j u_{23(jk)} = \sum_k u_{23(jk)} = 0, \\ &\sum_i u_{123(ijk)} = \sum_j u_{123(ijk)} = \sum_k u_{123(ijk)} = 0. \end{aligned} \tag{3.12}$$

This general model imposes no restrictions on the $\{m_{ijk}\}$ and corresponds to model (4) above. Note that the model given by formulas (3.11) and (3.12) is expressed in such a way that it describes the expected values for an $I \times J \times K$ table and not just the special case of the $2 \times 2 \times 2$ table of our lizard example.

We have already seen that setting

$$u_{12(ij)} = u_{13(ik)} = u_{23(jk)} = u_{123(ijk)} = 0$$

for all $i, j, k$ in expression (3.11) yields the model of complete independence of the three underlying variables. Before going on to look at other special cases of the loglinear model corresponding to models (1) through (3), let us stop for a moment and consider various ways to generate three-dimensional tables of counts.

## 3.2 Sampling Models

For all of the loglinear models we shall consider, we obtain the same maximum-likelihood estimates (MLEs) for the expected cell counts under a variety of different sampling distributions. In particular, the three sampling models described in Section 2.4 yield the same MLEs. A more technical description of these sampling models and their interrelationships follows.

The simplest of these sampling models results from a procedure in which the $IJK$ observed cell counts $\{x_{ijk}\}$ are viewed as having independent Poisson distributions with the expected counts $\{m_{ijk}\}$ as their means. Thus the probability function for this model is

$$\prod_{i,j,k} \frac{m_{ijk}{}^{x_{ijk}} e^{-m_{ijk}}}{x_{ijk}!}. \tag{3.13}$$

Since the cells contain counts having independent Poisson distributions, the total count in the table, $x_{+++}$, has a Poisson distribution with mean $m_{+++} = \Sigma_{i,j,k}\, m_{ijk}$. This fact is used below.

The second sampling procedure is the *multinomial* one in which a sample of $N$ individuals or objects is cross-classified according to the categories of the three variables. The probability function for the multinomial model is

$$\frac{N!}{\Pi_{i,j,k}\, x_{ijk}!} \prod_{i,j,k} p_{ijk}{}^{x_{ijk}}, \tag{3.14}$$

where $\Sigma_{i,j,k}\, p_{ijk} = 1$. The expected or mean values of the $\{x_{ijk}\}$ are $\{N p_{ijk}\}$.

We note that if the $\{x_{ijk}\}$ have been generated according to a Poisson sampling scheme, then the conditional distribution of the $\{x_{ijk}\}$, given that $x_{+++} = N$, is multinomial with probability function (3.14), where $p_{ijk} = m_{ijk}/m_{+++}$.

The third sampling model, which we refer to as *product-multinomial*, typically occurs in situations where one or more of the variables can be thought of as explanatory variables, and the remainder as response variables. Then for each combination of the explanatory variables we pick a fixed sample size, and the cross-classification of each sample according to the response variables is governed by a multinomial distribution. For example, suppose we fix the $K$ layer totals, $\{x_{++k}\}$, and let $P_{ijk}$ be the probability of an observation falling into the $i$th category of variable 1 and the $j$th category of variable 2, given that it falls into the $k$th category of variable 3 (i.e., $P_{ijk} = p_{ijk}/p_{++k}$). Then the probability function of this product-multinomial sampling scheme is

$$\prod_{k} \left[ \frac{x_{++k}!}{\Pi_{i,j}\, x_{ijk}!} \prod_{i,j} P_{ijk}{}^{x_{ijk}} \right]. \tag{3.15}$$

Thus we have independent multinomial samples for the $K$ layers. The expected values of the $\{x_{ijk}\}$ for this model are $\{x_{++k} P_{ijk}\}$. Similarly we could fix a set of two-dimensional marginal totals, say $\{x_{+jk}\}$, and have a product-multinomial scheme with independent multinomial samples for each column-by-layer combination.

We note that the product-multinomial probability function given by expression (3.15) is also the conditional probability function for $\{x_{ijk}\}$ generated by a multinomial sampling scheme, given the $\{x_{++k}\}$.

Birch [1963], Haberman [1974a], and others have shown that the MLEs of the expected cell values under the loglinear models we are about to consider are the same under all three sampling schemes discussed above. The one condition required for this result is that the $u$-terms "corresponding" to the fixed margins in the product-multinomial sampling model be included in the loglinear model under consideration. We elaborate on this point in Chapter 6.

In Chapters 1 and 2 it was noted that, in addition to considering the sampling model, we should try to distinguish between response and explanatory variables (see also Bhapkar and Koch [1968]). For example, in the lizard data of Table 3-1, species can be thought of as an explanatory variable or factor, and perch height and perch diameter as response variables. More generally, for three-dimensional contingency tables there are three situations among which we should distinguish:

(i) three-response, no factor;
(ii) two-response, one-factor;
(iii) one-response, two-factor.

For situation (i) only Poisson or multinomial sampling models are appropriate, whereas for situations (ii) and (iii) we could also use a product-multinomial model in which the fixed marginal totals correspond to explanatory variables or factors.

A sensible approach for the analysis of data with one or two explanatory variables is to condition on the values of these factors, treating them as fixed even in those cases where they are not. Such an approach is discussed in more detail when we consider logit models in Chapter 6.

There are some situations, such as case-control studies in epidemiology, in which the marginal totals corresponding to the response variable (case or control) is fixed. Either one can condition on the explanatory variables, obtaining a generalized hypergeometric sampling model, or one can treat one of the explanatory variables as if it were the response. We discuss the rationale for the latter approach in the Section 7.5.

### 3.3 Estimated Expected Values

We now return to the discussion of the form of estimated expected cell values for various loglinear models. We have already introduced formula (3.5) for the estimated expected values under the model of complete independence:

$$\hat{m}_{ijk} = \frac{x_{i++}x_{+j+}x_{++k}}{N^2}. \tag{3.16}$$

These are maximum-likelihood estimates when the sampling model is Poisson, multinomial, or product-multinomial with one set of one-dimensional marginal totals fixed.

Now if we take the general loglinear model in (3.11) and set

$$u_{12(ij)} = u_{123(ijk)} = 0 \tag{3.17}$$

for all $i$, $j$, $k$, then

$$m_{+jk} = e^{u+u_{2(j)}+u_{3(k)}+u_{23(jk)}} \sum_i e^{u_{1(i)}+u_{13(ik)}}, \tag{3.18}$$

$$m_{i+k} = e^{u+u_{1(i)}+u_{3(k)}+u_{13(ik)}} \sum_j e^{u_{2(j)}+u_{23(jk)}}, \tag{3.19}$$

and

$$m_{++k} = e^{u+u_{3(k)}} \sum_{i,j} e^{u_{1(i)}+u_{2(j)}+u_{13(ik)}+u_{23(jk)}}. \tag{3.20}$$

Dividing the product of the right-hand sides of expressions (3.18) and (3.19) by the right-hand side of expression (3.20) yields

$$m_{ijk} = \frac{m_{i+k}m_{+jk}}{m_{++k}}. \tag{3.21}$$

This is model (2) from Section 3.1, and it implies that variables 1 and 2 are independent for each fixed value of variable 3 (e.g., for the data in Table 3-1, perch height and perch diameter are independent, given species). For each fixed value of $k$ we get independence in the corresponding $I \times J$ subtable. This model is appropriate for consideration when the sampling is Poisson, multinomial, or product-multinomial with one set of fixed marginal totals, as long as the fixed totals are not $\{x_{ij+}\}$. Looking at the likelihood or probability functions for these sampling models, we find that $\{x_{i+k}\}$ and $\{x_{+jk}\}$ are complete minimal sufficient statistics* for $\{m_{i+k}\}$ and $\{m_{+jk}\}$. Maximizing the likelihood function in any of these cases leads to replacing expected marginal totals in expression (3.21) by the corresponding observed totals:

*For a nontechnical discussion of minimal sufficient statistics and other mathematical statistics terminology, see Appendix I.

$$\hat{m}_{ijk} = \frac{x_{i+k}x_{+jk}}{x_{++k}} \tag{3.22}$$

(for details see Birch [1963] or Bishop, Fienberg, and Holland [1975]).
If we set

$$u_{13(ik)} = u_{23(jk)} = u_{123(ijk)} = 0 \tag{3.23}$$

for all $i, j, k$, then we have a model that is appropriate for consideration when the sampling model is Poisson, multinomial, or product-multinomial with the set of fixed marginal totals corresponding to variable 1, 2, or 3, or to variables 1 and 2 jointly. Some algebraic manipulations similar to those for the conditional-independence model above yield

$$m_{ijk} = \frac{m_{ij+}m_{++k}}{m_{+++}}. \tag{3.24}$$

This is model (1) from Section 3.1, and it implies that variables 1 and 2, taken jointly, are independent of variable 3 (e.g., for the data in Table 3-1, perch height and perch diameter are jointly independent of species). If we think of the first two subscripts as one, then the expected values have the same form as in two-dimensional tables. Here $\{x_{ij+}\}$ and $\{x_{++k}\}$ are complete minimal sufficient statistics for $\{m_{ij+}\}$ and $\{m_{++k}\}$, and the estimated expected values are of the form

$$\hat{m}_{ijk} = \frac{x_{ij+}x_{++k}}{N}. \tag{3.25}$$

Finally, the remaining model, which is usually referred to as the *no second-order interaction model*, corresponds to setting

$$u_{123(ijk)} = 0 \tag{3.26}$$

for all $i, j, k$. This is model (3) of Section 3.1, and it can be considered in conjunction with Poisson, multinomial, or product-multinomial sampling models when either a set of one-dimensional or a set of two-dimensional marginal totals is fixed. The complete minimal sufficient statistics now are $\{x_{ij+}\}$, $\{x_{i+k}\}$, and $\{x_{+jk}\}$. We cannot write $m_{ijk}$ as a closed-form expression involving the marginal totals $\{m_{ij+}\}$, $\{m_{+jk}\}$, and $\{m_{i+k}\}$ as we did for the other four models; for example,

$$m_{ijk} \neq \left(\frac{m_{ij+}m_{i+k}m_{+jk}}{m_{i++}m_{+j+}m_{++k}}\right)m_{+++}. \tag{3.27}$$

The estimated expected values for this model are, however, functions of the three sets of two-dimensional marginal totals,

$$\{x_{ij+}\}, \quad \{x_{+jk}\}, \quad \{x_{i+k}\}. \tag{3.28}$$

In fact we have the following general rule for getting the estimated expected values for all five models:

(1) for each variable, look at the highest-order effect in the loglinear model involving that variable;

(2) compute the observed marginal totals corresponding to the highest-order effects in (1)—e.g., $\{x_{ij+} \mid i = 1, 2; j = 1, 2\}$ corresponds to $\{u_{12(ij)} \mid i = 1, 2; j = 1, 2\}$;

(3) estimate the expected values for the model using only the sets of observed marginal totals in (2), or totals that can be computed from them.

This rule follows from the general results described in Appendix II. We shall consider the computation of expected values under the no second-order interaction model in detail shortly.

Going back to the $2 \times 2 \times 2$ example in Table 3-1, we already have considered model (2) since we have checked for independence between perch height and perch diameter for each of the two species separately, and the expected values under this model are just those given in Tables 2-2b and 2-3b and reproduced in column (4) of Table 3-1. Since this model of conditional independence fits so well, it makes little practical sense to look at more complicated models based on additional parameters, though we shall do so later for illustrative purposes. Turning to the model of complete independence of the three variables, we find the estimated expected values, computed using (3.16), displayed in column (3) of Table 3-1. The fit of these expected values to the observed counts is not as good as it was in previous examples. In particular, note the discrepancies in the (2, 1, 1), (1, 2, 1), and (2, 1, 2) cells. We consider formal methods for examining the goodness-of-fit of this model in Section 3.5.

Table 3-2a contains further data on the structural habitat of *Anolis* lizards of Bimini, this time for *sagrei* adult males and *angusticeps* adult males. This table is also based on data reported in Schoener [1968]. Note that the class boundaries of the dichotomies for perch height and perch diameter here are not the same as in Tables 2-2a and 2-3a. The observed cross-product ratios for the $2 \times 2$ tables corresponding to the two species are $\hat{\alpha}_1 = 1.46$ (*sagrei*) and $\hat{\alpha}_2 = 14.0$ (*angusticeps*). These appear to be unequal and clearly different from 1, but they are subject to sampling variability so we cannot be sure.

We could begin an analysis of these data by considering the model that specifies equality of the cross-product ratios in each level—i.e., $\alpha_1 = \alpha_2$—but does not specify the common value

**Table 3-2**
Counts for Structural Habitat Categories for *Anolis* Lizards of Bimini: *sagrei* Adult Males vs. *angusticeps* Adult Males (Schoener [1968])

(a) Observed data

| | | perch diameter (inches) | | | |
|---|---|---|---|---|---|
| | | *sagrei* | | *angusticeps* | |
| | | ≤2.5 | >2.5 | ≤2.5 | >2.5 |
| perch height (feet) | > 5.0 | 15 | 18 | 21 | 1 |
| | ≤ 5.0 | 48 | 84 | 3 | 2 |

(b) 2 × 2 marginal total

| | | perch diameter | |
|---|---|---|---|
| | | ≤2.5 | >2.5 |
| perch height | > 5.0 | 36 | 19 |
| | ≤ 5.0 | 51 | 86 |

(c) 2 × 2 marginal total

| | | *sagrei* | *angusticeps* |
|---|---|---|---|
| perch height | > 5.0 | 33 | 22 |
| | ≤ 5.0 | 132 | 5 |

(d) 2 × 2 marginal total

| | | *sagrei* | *angusticeps* |
|---|---|---|---|
| perch diameter | ≤ 2.5 | 63 | 24 |
| | > 2.5 | 102 | 3 |

$$\frac{m_{111}m_{221}}{m_{121}m_{211}} = \frac{m_{112}m_{222}}{m_{122}m_{212}}. \tag{3.29}$$

Setting $p_{ijk} = m_{ijk}/N$, where $m_{+++} = N$, we find that (3.29) is equivalent to the model

$$\frac{p_{111}p_{221}}{p_{121}p_{211}} = \frac{p_{112}p_{222}}{p_{122}p_{212}}. \tag{3.30}$$

The loglinear model that postulates no second-order interaction ($u_{123(ijk)} = 0$ for all $i, j, k$) is equivalent to the model specified by expression (3.29) and was first considered by Bartlett [1935]. One way to test $u_{123} = 0$ in a $2 \times 2 \times 2$ table is to consider the test statistic

$$z = \frac{\log \hat{\alpha}_1 - \log \hat{\alpha}_2}{\sqrt{s_{\hat{\alpha}_1}^2 + s_{\hat{\alpha}_2}^2}}, \tag{3.31}$$

where $s_{\hat{\alpha}_i}$ is the estimated large-sample standard deviation of $\log \hat{\alpha}_i$ given by (2.29). If $\alpha_1 = \alpha_2$ (i.e., $u_{123} = 0$), then this statistic has an asymptotic normal distribution with mean zero and variance one. The square of $z$, which has an asymptotic $\chi^2$ distribution with 1 d.f., has been used by many authors; for example, $z^2$ is a special case of the minimum modified chi-square statistic discussed by Grizzle, Starmer, and Koch [1969].

## 3.4 Iterative Computation of Expected Values

A second way to test the hypothesis $u_{123} = 0$ in the example above is to compute the estimated expected cell values for this model. By the rules described above, the $\{\hat{m}_{ijk}\}$ are functions only of $\{x_{ij+}\}$, $\{x_{i+k}\}$, and $\{x_{+jk}\}$. Using the method of maximum likelihood (see Appendix II) we find that the $\{\hat{m}_{ijk}\}$ must satisfy:

$$\begin{aligned} \hat{m}_{ij+} &= x_{ij+}, \qquad i,j = 1,2, \\ \hat{m}_{i+k} &= x_{i+k}, \qquad i,k = 1,2, \\ \hat{m}_{+jk} &= x_{+jk}, \qquad j,k = 1,2. \end{aligned} \tag{3.32}$$

Equations (3.29) and (3.32) uniquely define a set of estimated expected values. Unfortunately we cannot write these out in closed form. The following iterative procedure, however, yields the MLEs:

**Step 1:** Set

$$\hat{m}_{ijk}^{(0)} = 1 \quad \text{for all } i,j,k. \tag{3.33}$$

Then for $v = 0$ compute

**Step 2:**

$$\hat{m}_{ijk}^{(3v+1)} = \frac{x_{ij+}}{\hat{m}_{ij+}^{(3v)}} \hat{m}_{ijk}^{(3v)}, \tag{3.34}$$

**Step 3:**

$$\hat{m}_{ijk}^{(3v+2)} = \frac{x_{i+k}}{\hat{m}_{i+k}^{(3v+1)}} \hat{m}_{ijk}^{(3v+1)}, \tag{3.35}$$

**Step 4:**

$$\hat{m}_{ijk}^{(3(v+1))} = \frac{x_{+jk}}{\hat{m}_{+jk}^{(3v+2)}} \hat{m}_{ijk}^{(3v+2)}. \tag{3.36}$$

This completes the first cycle of the iteration. Step 2 makes the estimated expected values satisfy the first set of marginal constraints in (3.32). Step 3 makes them satisfy the second set—$\hat{m}_{i+k}^{(2)} = x_{i+k}$ for all $i$ and $k$—as well as the first set of constraints—$\hat{m}_{ij+}^{(2)} = x_{ij+}$ for all $i$ and $j$. Similarly, Step 4 makes the expected values satisfy $\hat{m}_{+jk}^{(3)} = x_{+jk}$ for all $j$ and $k$, but now the other two sets of marginal constraints are messed up—$\hat{m}_{ij+}^{(3)} \neq x_{ij+}$ and $\hat{m}_{i+k}^{(3)} \neq x_{i+k}$.

Repeating the cycle (3.34)–(3.36) for $v = 1, 2, \ldots$ yields an iterative procedure that ultimately converges to the estimated expected values $\{\hat{m}_{ijk}\}$. If the change in the expected values from one cycle to the next is sufficiently small, we terminate the iteration; otherwise we perform a further cycle.

At Step 1 of the iteration we took as starting values the vector of ones. Any set of starting values satisfying the no second-order interaction model, $u_{123} = 0$, would do, but Bishop, Fienberg, and Holland [1975] show that the speed of convergence of the iterative method is not substantially improved by the use of different starting values. Moreover, unit starting values are also appropriate for other loglinear models, and their use simplifies the task of preparing computer programs. General-purpose computer programs that perform the iterative procedure just described are available at a large number of computer installations and are part of several widely used program packages. Detailed Fortran listings are available in Bishop [1967, Appendix I] and Haberman [1972, 1973b].

The iterative proportional fitting procedure was first introduced for work with census data by Deming and Stephan [1940a, b], and different proofs of convergence have been given by Brown [1959], Ireland and Kullback [1968], Fienberg [1970b], and others. An interesting generalization is provided by Gokhale [1971] and by Darroch and Ratcliff [1972].

To apply the iterative procedure to the data in Table 3-2, we begin with initial cell values equal to 1, that is, $\hat{m}_{ijk}^{(0)} = 1$ for all $i, j, k$. To get the values at the end of Step 2 we add the initial values over variable 3: for example,

$$\hat{m}_{11+}^{(0)} = \hat{m}_{111}^{(0)} + \hat{m}_{112}^{(0)} = 2.$$

Then we multiply each of the intital values forming this sum by the ratio $x_{11+}/\hat{m}_{11+}^{(0)}$, yielding

$$\hat{m}_{111}^{(1)} = \hat{m}_{112}^{(1)} = 1 \times 36/2 = 18.0.$$

The other $\hat{m}_{ijk}^{(1)}$ are computed in a similar fashion. For Step 3 we add the $\hat{m}_{ijk}^{(1)}$ over variable 2: for example,

$$\hat{m}_{1+1}^{(1)} = \hat{m}_{111}^{(1)} + \hat{m}_{121}^{(1)} = 18.0 + 9.5 = 27.5.$$

Our estimates at this step are of the form

$$\hat{m}_{111}^{(2)} = \hat{m}_{111}^{(1)} \times (x_{1+1}/\hat{m}_{1+1}^{(1)}) = 18 \times 33/27.5 = 21.6.$$

**Table 3-3**
Estimated Expected Values for Entries in Table 3-2, Based on the Model of All Two-Factor Effects but No Three-Factor Effect, and Cell Values at Various Stages of the Iterative Procedure.

| Cell $(i, j, k)$ | $\hat{m}_{ijk}^{(0)}$ | $\hat{m}_{ijk}^{(1)}$ | $\hat{m}_{ijk}^{(2)}$ | $\hat{m}_{ijk}^{(3)}$ | $\hat{m}_{ijk}^{(6)}$ | $\hat{m}_{ijk}^{(9)}$ | $\hat{m}_{ijk}^{(12)} = \hat{m}_{ijk}$ |
|---|---|---|---|---|---|---|---|
| (1, 1, 1) | 1.0 | 18.0 | 21.6 | 19.2 | 16.5 | 16.2 | 16.2 |
| (2, 1, 1) | 1.0 | 25.5 | 49.1 | 43.8 | 46.5 | 46.8 | 46.8 |
| (1, 2, 1) | 1.0 | 9.5 | 11.4 | 12.3 | 16.3 | 16.7 | 16.8 |
| (2, 2, 1) | 1.0 | 43.0 | 82.9 | 89.7 | 85.7 | 85.3 | 85.2 |
| (1, 1, 2) | 1.0 | 18.0 | 14.4 | 21.3 | 19.9 | 19.8 | 19.8 |
| (2, 1, 2) | 1.0 | 25.5 | 1.9 | 2.7 | 4.1 | 4.2 | 4.2 |
| (1, 2, 2) | 1.0 | 9.5 | 7.6 | 2.1 | 2.2 | 2.2 | 2.2 |
| (2, 2, 2) | 1.0 | 43.0 | 3.1 | 0.9 | 0.8 | 0.8 | 0.8 |

Table 3-3 contains the estimated expected cell values after each step in the first cycle, and then after each subsequent cycle of the iteration. At the end of the fourth cycle the estimated expected values differed from those at the end of the preceding cycle by less than 0.1, and the iteration was terminated.

An alternate but essentially equivalent criterion for termination of the iterative procedure would be to stop when the three sets of expected two-dimensional marginal totals differed by 0.1 or less from the observed marginal totals. Bishop, Fienberg, and Holland [1975] provide a detailed discussion of this point.

With only a slight modification, the iterative method can also be used to compute estimated expected values for the other models. As an example, consider the model corresponding to the independence of variables 2 and 3 conditional on the level of variable 1. In this model each cycle of the iterative method would have only two steps. The first adjusts for the marginal totals $\{x_{ij+}\}$, and the second for the totals $\{x_{i+k}\}$. The third set of marginal totals used earlier is not needed here according to our general rule. Suppose the table is of size $2 \times 2 \times 2$. Then $\hat{m}_{ijk}^{(0)} = 1$ and

$$\hat{m}_{ijk}^{(1)} = \frac{x_{ij+}}{2}. \tag{3.37}$$

Adding $\hat{m}_{ijk}^{(1)}$ over variable 2 yields

$$\hat{m}_{i+k}^{(1)} = \frac{x_{i1+}}{2} + \frac{x_{i2+}}{2} = \frac{x_{i++}}{2}, \tag{3.38}$$

and so

$$\hat{m}_{ijk}^{(2)} = \frac{x_{ij+}}{2} \times \frac{x_{i+k}}{(x_{i++}/2)} = \frac{x_{ij+}x_{i+k}}{x_{i++}}, \tag{3.39}$$

the estimated expected value for the model of conditional independence. Only one cycle of the iteration is needed for this model, which we could have fitted directly, and this cycle simply gives a systematic way for carrying out the direct computation. More generally, in tables with no more than six dimensions, when a model can be fitted directly, one cycle of the iteration will produce exactly the same cell estimates. In seven-dimensional tables a second cycle is sometimes necessary for particular orderings of the steps within the cycle. At any rate, one can always use the iterative method for all models of the class discussed above without worrying about whether direct estimation is possible.

## 3.5 Goodness-of-Fit Statistics

Once we have estimated expected values under one of the loglinear models, we can check the goodness-of-fit of the model using either of the following statistics:

$$X^2 = \sum \frac{(\text{Observed} - \text{Expected})^2}{\text{Expected}}, \tag{3.40}$$

$$G^2 = 2\sum (\text{Observed}) \log \left(\frac{\text{Observed}}{\text{Expected}}\right), \tag{3.41}$$

where the summation in both cases is over all cells in the table. Expression (3.40) is the general form of the Pearson chi-square statistic, and expression (3.41) is $-2$ times the logarithm of the likelihood-ratio test statistic used for testing that the model fitted is correct versus the unrestricted alternative (for more details see Rao [1973], p. 417). If the model fitted is correct and the total sample size is large, both $X^2$ and $G^2$ have approximate $\chi^2$ distributions with degrees of freedom given by the following formula:

$$\text{d.f.} = \#\ \text{cells} - \#\ \text{parameters fitted}. \tag{3.42}$$

Applying this formula to each of our five models we get the degrees of freedom indicated in Table 3-4 (the values in the last column in square brackets are for an $I \times J \times K$ table). In the abbreviated bracket notation in column 2 of Table 3-4 we describe a model by means of the highest order $u$-terms present; for example, [12] means that $u_{12}$ is present in the model. This notation uniquely describes all hierarchical loglinear models.

The statistics $X^2$ and $G^2$ are asymptotically equivalent, that is, they are

**Table 3-4**
Degrees of Freedom Associated with Various Loglinear Models for Three-Dimensional Tables

| Model | Abbreviation | # parameters fitted* | d.f.* |
|---|---|---|---|
| $u + u_1 + u_2 + u_3$ | [1][2][3] | 4<br>$[1 + (I-1) + (J-1) + (K-1)]$ | 4<br>$[IJK - I - J - K + 2]$ |
| $u + u_1 + u_2 + u_3 + u_{12}$ | [12][3] | 5<br>$[1 + (I-1) + (J-1) + (K-1) + (I-1)(J-1)]$ | 3<br>$[(K-1)(IJ-1)]$ |
| $u + u_1 + u_2 + u_3 + u_{12} + u_{23}$ | [12][23] | 6<br>$[1 + (I-1) + (J-1) + (K-1) + (I-1)(J-1) + (J-1)(K-1)]$ | 2<br>$[J(I-1)(K-1)]$ |
| $u + u_1 + u_2 + u_3 + u_{12} + u_{23} + u_{13}$ | [12][23][13] | 7<br>$[1 + (I-1) + (J-1) + (K-1) + (I-1)(J-1) + (J-1)(K-1) + (I-1)(K-1)]$ | 1<br>$[(I-1)(J-1)(K-1)]$ |
| $u + u_1 + u_2 + u_3 + u_{12} + u_{23} + u_{13} + u_{123}$ | [123] | 8<br>$IJK$ | 0 |

*The first entry pertains to the $2 \times 2 \times 2$ table. The second entry pertains to the $I \times J \times K$ table.

equivalent in very large samples when the null hypothesis is true. For our purposes we interpret "very large" to mean that the total sample size is at least ten times the number of cells in the table. This is, of course, a very rough rule of thumb, and we use $X^2$ and $G^2$ even when the sample size is much smaller. Issues with regard to the small-sample adequacy of the $\chi^2$ approximations to $X^2$ and $G^2$ and with regard to the statistical power of the two test statistics for various composite alternatives have not yet been completely resolved (see Larntz [1973]).

Table 3-5 contains a summary of the values of $G^2$ and $X^2$ for each of the eight possible loglinear models as applied to the data in Tables 3-1 and 3-2. We note that the values of $X^2$ and $G^2$ are usually quite close, except when both are large, and are far out in the tail of the corresponding $\chi^2$ distributions. For example, for the data in Table 3-1 and the model of complete independence of the three variables, the statistics are

$$X^2 = \frac{(32-35.8)^2}{35.8} + \frac{(86-65.2)^2}{65.2} + \frac{(11-22.3)^2}{22.3} + \frac{(35-40.6)^2}{40.6}$$
$$+ \frac{(61-53.5)^2}{53.5} + \frac{(73-97.4)^2}{97.4} + \frac{(41-33.3)^2}{33.3} + \frac{(70-60.7)^2}{60.7}$$
$$= 23.91$$

and

$$G^2 = 64 \log(32/35.8) + 172 \log(86/65.2) + 22 \log(11/22.3)$$
$$+ 70 \log(35/40.6) + 122 \log(61/53.5) + 146 \log(73/97.4)$$
$$+ 82 \log(41/33.3) + 140 \log(70/60.7)$$
$$= 25.42.$$

The value of $G^2$ computed here is slightly larger than the one in Table 3-5 because the estimated expected values used in the computation above were rounded to one decimal place. The two statistics here have almost the same value. The number of degrees of freedom associated with each of them is 4, and thus either the Pearson or the likelihood-ratio value is significant at the 0.001 level when referred to a chi-square table on 4 d.f. (the 0.001 value in the table is 18.47). This result implies that the model does not provide an adequate fit to the data.

Clearly the only model that provides a good fit to both sets of data is the no three-factor interaction model: $u_{123} = 0$. We have already seen that the model specifying conditional independence of perch height and perch

**Table 3-5**
Loglinear Models Fit to Data in Tables 3-1 and 3-2, and Their Corresponding Goodness-of-Fit Statistics

| Model | Data from Table 3-1 | | Data from Table 3-2 | | |
|---|---|---|---|---|---|
| | $X^2$ | $G^2$ | $X^2$ | $G^2$ | d.f. |
| [1] [2] [3] | 23.91* | 25.04* | 111.30* | 70.08* | 4 |
| [1] [23] | 12.30* | 12.43* | 47.46* | 43.87* | 3 |
| [12] [3] | 23.98* | 24.43* | 72.23* | 57.39* | 3 |
| [13] [2] | 14.45* | 14.63* | 27.02* | 31.09* | 3 |
| [13] [23] | 2.02 | 2.03 | 6.11* | 4.88 | 2 |
| [12] [23] | 11.62* | 11.82* | 29.55* | 31.18* | 2 |
| [12] [13] | 13.78* | 14.02* | 15.75* | 18.40* | 2 |
| [12] [13] [23] | 0.15 | 0.15 | 2.71 | 3.02 | 1 |

*The asterisk indicates goodness-of-fit statistic values in the upper 5% tail of the corresponding $\chi^2$ distribution, with d.f. as indicated.

diameter given species fits the data of Table 3-1 quite well. (The $X^2$ value in Table 3-5 is just the sum of the two $X^2$ values computed for the two $2 \times 2$ tables separately.) For the data of Table 3-2, however, there is some question as to whether this conditional-independence model fits, since $X^2$ exceeds the 0.05 significance value while $G^2$ does not. The problem of model selection for these data will be discussed further in Chapter 4.

### 3.6 Hierarchical Models

We have not considered all possible variants of our loglinear model. For example, we have not considered the model

$$\log m_{ijk} = u + u_{1(i)} + u_{2(j)} + u_{3(k)} + u_{123(ijk)}, \tag{3.43}$$

subject to the usual ANOVA-like constraints. In Section 3.1 we showed how successively higher-order $u$-terms measure deviations from lower-order terms (see, for example, expression (3.8)). In order to retain this interpretation we limit our models to a hierarchical set in which higher-order terms may be included only if the related lower-order terms are included. Thus $u_{123}$ cannot be included in a model unless $u_{12}$, $u_{13}$, and $u_{23}$ are all in the model. Clearly (3.43) is not a hierarchical model in this sense. It is possible to consider fitting nonhierarchical models such as (3.43) to data, but we cannot then compute the estimated expected values directly via our iterative proportional fitting procedure. Rather, we need to transform the table, interchanging cells, so that the nonhierarchical model for the original table becomes a hierarchical model for the transformed table. Such transformations are quite straightforward for $2^k$ tables (for details see Bloomfield [1974]). Haberman [1974a] gives some general results that can be used directly to get likelihood equations for nonhierarchical loglinear models.

Our primary reason for avoiding nonhierarchical models is interpretive rather than technical. The basic feature of ANOVA-like models is what Nelder [1976] refers to as *marginality*. For example, for two sets of main effects $\{u_{1(i)}\}$ and $\{u_{2(j)}\}$ and an associated set of interaction effects $\{u_{12(ij)}\}$, the main effects are marginal to the interaction effects. This can be seen from the restrictions placed on the interaction term (especially geometrically); that is, if one looks at interaction terms directly in terms of contrasts, they include the contrasts for the main effects. More importantly, if the interaction terms are nonnegligible, then it is of little interest that the main effects may be zero. The same remarks apply to higher-order interactions and the marginality of lower-order ones (i.e., $u_{12}$ is marginal to $u_{123}$).

### 3.7 A Further Example

In Section 2.4 we examined the three two-dimensional marginal tables associated with a three-dimensional cross-classification. Here we look at the actual three-dimensional table for this example and compare a loglinear model analysis of this table with the earlier analyses.

The data involved in this example refer to a population of World War II volunteers for the pilot, navigator, and bombardier training programs of the U.S. Army Air Corps. Of the approximately 500,000 volunteers, roughly 75,000 passed the qualifying examination at the required level and then took a battery of 17 aptitude-like tests during the period July–December 1943. Then, in the 1950s, Thorndike and Hagen [1959] undertook a study of a random sample of 17,000 of the group of 75,000 individuals to explore the predictive ability of the tests with regard to vocational success. Of these 17,000 about 2000 were still in the military as of 1955, and 1500 had died since 1943. Thorndike and Hagen received detailed responses from 9700 of the remaining 13,500, a response rate of 72 percent. (This is a high response rate for this sort of survey.)

In 1969 the National Bureau of Economic Research did a follow-up survey of the 7500 of Thorndike and Hagen's 9700 civilian respondents for whom up-to-date addresses were available. About 70 percent of these people eventually answered the NBER questionnaire, and the data in Table 3-6 are based on 4353, or approximately 83 percent, of the NBER respondents. We have given extensive background for this example because it will be extremely important when we attempt to draw conclusions from our analysis. In particular, we note that the 1969 NBER sample tended to be better-educated and to have higher ability than the 1955 sample. Taubman and Wales [1974] give further details.

Table 3-6 presents the full $4 \times 4 \times 5$ cross-classification of the 4353 individuals from the NBER–TH sample by occupation, education, and aptitude. Individuals are classified into one of five ability groupings based on the 1943 tests of reading comprehension, numerical operations, and mathematics. The A5 group scored in the upper quintile on at least two of the three tests but had no score in the lowest quintile. The A4 group had one upper-quintile score and no lower-quintile one. Groups A1 and A2 are similar to A5 and A4, respectively, with the lowest and highest quintiles reversed, and A3 contains all other possible scores. Almost all respondents had completed high school (i.e., 12 years of education) before applying for aviation cadet training, and many of those with college and graduate training received this education following World War II with financial support resulting from the

**Table 3-6**
Three-Dimensional Cross-Classification of 4353 Individuals into 4 Occupational Groups (O), 4 Educational Levels (E), and 5 Aptitude Levels (A) as Measured by a Scholastic Aptitude Test (Beaton [1975])

| O1 | (Self-employed, business) | | | | |
|---|---|---|---|---|---|
| | E1 | E2 | E3 | E4 | Totals |
| A1 | 42 | 55 | 22 | 3 | 122 |
| A2 | 72 | 82 | 60 | 12 | 226 |
| A3 | 90 | 106 | 85 | 25 | 306 |
| A4 | 27 | 48 | 47 | 8 | 130 |
| A5 | 8 | 18 | 19 | 5 | 50 |
| Totals | 239 | 309 | 233 | 53 | 834 |

| O2 | (Self-employed, professional) | | | | |
|---|---|---|---|---|---|
| | E1 | E2 | E3 | E4 | Totals |
| A1 | 1 | 2 | 8 | 19 | 30 |
| A2 | 1 | 2 | 15 | 33 | 51 |
| A3 | 2 | 5 | 25 | 83 | 115 |
| A4 | 2 | 2 | 10 | 45 | 59 |
| A5 | 0 | 0 | 12 | 19 | 31 |
| Totals | 6 | 11 | 70 | 199 | 286 |

| O3 | (Teacher) | | | | |
|---|---|---|---|---|---|
| | E1 | E2 | E3 | E4 | Totals |
| A1 | 0 | 0 | 1 | 19 | 20 |
| A2 | 0 | 3 | 3 | 60 | 63 |
| A3 | 1 | 4 | 5 | 86 | 96 |
| A4 | 0 | 0 | 2 | 36 | 38 |
| A5 | 0 | 0 | 1 | 14 | 15 |
| Totals | 1 | 7 | 12 | 215 | 235 |

| O4 | (Salary-employed) | | | | |
|---|---|---|---|---|---|
| | E1 | E2 | E3 | E4 | Totals |
| A1 | 172 | 151 | 107 | 42 | 472 |
| A2 | 208 | 198 | 206 | 92 | 704 |
| A3 | 279 | 271 | 331 | 191 | 1072 |
| A4 | 99 | 126 | 179 | 97 | 501 |
| A5 | 36 | 35 | 99 | 79 | 249 |
| Totals | 794 | 781 | 922 | 501 | 2998 |

GI Bill. Thus our measure of education is only partially antecedent to our measure of aptitude. Occupation is measured as of 1969.

In other analyses of the more complete information available for the NBER-TH sample, much attention was focused on the effects of education on attitudes and income. Since education appears to be related to income, one of the main reasons for analyzing the data in Table 3-6 is to answer the questions posed by Beaton [1975]: "Is education, in fact, correlated to ability, and does ability provide a possible alternative explanation for education effects found for other variables?"

If we ignore for the moment the various response biases present in the NBER–TH sample, a reasonable approach is to treat the data in Table 3-6 as having been generated via a multinomial sampling scheme. Despite the fact that aptitude and education are antecedents to occupation, we have chosen for the time being not to treat them as explanatory variables for the response variable, occupation. Table 3-7 gives the values of both goodness-of-fit statistics for each of the eight hierarchical models fitted to the three-dimensional data of Table 3-6. In this summary, variable 1 is occupation, variable 2 is aptitude, and variable 3 is education.

The two models that provide a reasonable fit to the data are

**Table 3-7**
Values of the Chi-Square Goodness-of-Fit Statistics, $X^2$ and $G^2$, for Various Log-linear Models as Applied to Data in Table 3-6

| Model | d.f. | $X^2$ | $G^2$ |
|---|---|---|---|
| [12] [13] [23] | 36 | 23.6 | 25.1 |
| [12] [13] | 48 | 184.6* | 190.8* |
| [13] [23] | 48 | 48.0 | 50.9 |
| [12] [23] | 45 | 1301.1* | 1142.2* |
| [12] [3] | 57 | 1424.1* | 1319.7* |
| [13] [2] | 60 | 226.7* | 228.2* |
| [23] [1] | 57 | 1336.8* | 1179.6* |
| [1] [2] [3] | 69 | 1519.8* | 1357.0* |

*Denotes values in upper 5% tail of the corresponding $\chi^2$ distribution, with d.f. as indicated.

(1) $u_{123(ijk)} = 0,$
(2) $u_{12(ij)} = u_{123(ijk)} = 0.$

Model (2) is a special case of (1) and implies that, given the level of education, the current occupational classification is independent of scholastic aptitude. One could not reach such a conclusion simply by looking at the two-dimensional marginal tables. The estimated expected cell values for this conditional-independence model, displayed in Table 3-8, are quite similar to the observed values in the sense that there are no glaring discrepancies, as indicated by the standardized residuals of the form $(x_{ijk} - \hat{m}_{ijk})/\hat{m}_{ijk}$.

Actually we must choose between models (1) and (2), and to do this we require some further theoretical results. Although we shall not complete the analysis of this example until after the discussion in Chapter 4, for the present we proceed as if model (2) is our model of choice. The next step in the analysis of these data should be to examine the estimated values of the interaction $u$-terms in model (2): $\{\hat{u}_{13(ik)}\}$ and $\{\hat{u}_{23(jk)}\}$. These estimates are given in Table 3-9. Occupation levels 1 and 4 (self-employed, business, and salary-employed) are positively related to low education and negatively related to high education with a gradual progression from one extreme to the other, while exactly the reverse seems to hold for teachers and self-employed professionals. This is as we would expect. There is also a natural progression across rows and down columns in the values of $\{\hat{u}_{23(jk)}\}$, with low education being positively related to low aptitude and high education positively related to high aptitude. These effects as displayed here account almost completely for the observed marginal relationship between aptitude and occupation. Tentative answers to the questions raised by Beaton would

**Table 3-8**
Estimated Expected Values under Model of Conditional Independence of Occupation and Aptitude Given Education

O1

| | E1 | E2 | E3 | E4 | Totals |
|---|---|---|---|---|---|
| A1 | 49.4 | 58.0 | 26.0 | 4.5 | 137.9 |
| A2 | 64.6 | 79.5 | 53.5 | 10.8 | 208.4 |
| A3 | 85.5 | 107.6 | 84.0 | 21.1 | 298.2 |
| A4 | 29.4 | 49.1 | 44.8 | 10.2 | 133.5 |
| A5 | 10.1 | 14.8 | 24.7 | 6.4 | 56.0 |
| Totals | 239.0 | 309.0 | 233.0 | 53.0 | 834.0 |

O2

| | E1 | E2 | E3 | E4 | Totals |
|---|---|---|---|---|---|
| A1 | 1.2 | 2.1 | 7.8 | 17.1 | 28.2 |
| A2 | 1.6 | 2.8 | 16.1 | 40.5 | 61.0 |
| A3 | 2.1 | 3.8 | 25.2 | 79.1 | 110.2 |
| A4 | 0.7 | 1.7 | 13.5 | 38.2 | 54.1 |
| A5 | 0.3 | 0.5 | 7.4 | 24.1 | 32.3 |
| Totals | 5.9 | 11.9 | 70.0 | 199.0 | 285.8 |

O3

| | E1 | E2 | E3 | E4 | Totals |
|---|---|---|---|---|---|
| A1 | 0.2 | 1.3 | 1.3 | 18.4 | 21.2 |
| A2 | 0.3 | 1.8 | 2.8 | 43.8 | 48.7 |
| A3 | 0.4 | 2.4 | 4.3 | 85.5 | 92.6 |
| A4 | 0.1 | 1.1 | 2.3 | 41.3 | 44.8 |
| A5 | 0.0 | 0.3 | 1.3 | 26.0 | 27.6 |
| Totals | 1.0 | 6.9 | 12.0 | 215.0 | 234.9 |

O4

| | E1 | E2 | E3 | E4 | Totals |
|---|---|---|---|---|---|
| A1 | 164.1 | 146.6 | 102.9 | 43.0 | 456.6 |
| A2 | 214.5 | 200.9 | 211.7 | 102.0 | 729.1 |
| A3 | 284.0 | 272.1 | 332.4 | 199.3 | 1087.8 |
| A4 | 97.7 | 124.1 | 177.4 | 96.3 | 495.5 |
| A5 | 33.6 | 37.4 | 97.6 | 60.6 | 229.2 |
| Totals | 793.9 | 781.1 | 922.0 | 501.2 | 2998.2 |

thus be that education *is* correlated with ability, in the way we would expect, and if we control for education, ability *is not* related to occupation, although it may still be related to other outcome variables such as salary.

One of the issues involved in comparing our earlier analysis of the NBER–

**Table 3-9**
Estimated Interaction $u$-Terms Corresponding to Estimated Expected Values in Table 3-8

(a) $\{\hat{u}_{13(ik)}\}$

| | E1 | E2 | E3 | E4 |
|---|---|---|---|---|
| O1 | 1.241 | 0.800 | −0.050 | −1.991 |
| O2 | −0.718 | −0.810 | 0.472 | 1.057 |
| O3 | −1.528 | −0.280 | −0.309 | 2.117 |
| O4 | 1.005 | 0.290 | −0.112 | −1.182 |

(b) $\{\hat{u}_{23(jk)}\}$

| | A1 | A2 | A3 | A4 | A5 |
|---|---|---|---|---|---|
| E1 | 0.460 | 0.186 | 0.040 | −0.225 | −0.461 |
| E2 | 0.323 | 0.095 | −0.028 | −0.011 | −0.379 |
| E3 | −0.275 | −0.096 | −0.071 | 0.103 | 0.338 |
| E4 | −0.507 | −0.185 | 0.058 | 0.133 | 0.501 |

TII data with the foregoing one is whether or not a test for interaction between variables 1 and 2 based on the two-dimensional marginal table is in fact a test for $u_{12} = 0$, where $u_{12}$ is the two-factor term from the loglinear model for the full three-dimensional array. This problem is taken up in the next section.

## 3.8 Collapsing Tables

Suppose we have a three-dimensional cross-classification and we are interested in drawing inferences about the two-factor interaction terms, $\{u_{12(ij)}\}$, in a loglinear model for the expected cell values $\{m_{ijk}\}$. If we collapse the cross-classification over variable 3, yielding the two-dimensional marginal table of counts $\{m_{ij+}\}$, we would like to know whether the two-factor interaction terms for this marginal table, $\{u^*_{12(ij)}\}$, are the same as the $\{u_{12(ij)}\}$ from the loglinear model for the three-dimensional table.

In order to appreciate the result that explains "when we can collapse," we first recall a well-known formula from standard multivariate statistical analysis regarding partial correlation coefficients (see, for example, Snedecor and Cochran [1967], p. 400):

$$\rho_{12\cdot 3} = \frac{\rho_{12} - \rho_{13}\rho_{23}}{\sqrt{(1-\rho_{13}^2)(1-\rho_{23}^2)}}, \tag{3.44}$$

where $\rho_{12\cdot 3}$ is the partial correlation between variables 1 and 2, controlling for variable 3, and $\rho_{12}$, $\rho_{23}$, and $\rho_{13}$ are the simple correlations. If $\rho_{13} = 0$ or $\rho_{23} = 0$, $\rho_{12\cdot 3}$ is a scalar multiple of the $\rho_{12}$, and we can test the hypothesis $\rho_{12\cdot 3} = 0$ by testing for $\rho_{12} = 0$.

In the general loglinear model (3.11) for the three-dimensional contingency table, we can think of the two-factor terms $\{u_{12(ij)}\}$ as being the two-factor interactions between variables 1 and 2, controlling for variable 3 (since the effects of variable 3 are taken into account in the model). By analogy with the correlation coefficient result above, the following seems quite reasonable.

**Theorem 3-1 In a three-dimensional table the interaction between two variables (as given by the appropriate $u$-terms) may be measured from the table of sums obtained by collapsing over the third variable if and only if the third variable is independent (again as given by the appropriate $u$-terms) of *at least one* of the two variables exhibiting the interaction.**

Bishop, Fienberg, and Holland [1975] give a detailed proof of this result as well as some natural generalizations.

We can illustrate the importance of this theorem in the context of our example of a cross-classification of a sample of individuals by occupation, education, and aptitude. The two models we have yet to choose between are

(1) $u_{123} = 0$,
(2) $u_{12} = u_{123} = 0$.

The theorem states that we can measure $u_{12}$ from the two-dimensional marginal table of occupation and education if and only if $u_{13} = 0$ or $u_{23} = 0$. Since neither $u_{13} = 0$ nor $u_{23} = 0$ in either of the models, we cannot choose between them on the basis of a test for no interaction carried out on the two-dimensional marginal table. Thus, as noted in Section 2.4, it is not sufficient to examine the three two-dimensional marginal tables resulting from the full cross-classification.

This theorem also provides a simple and straightforward explanation of the so-called Simpson's paradox (see Simpson [1951]). Blyth [1972] states the paradox as follows: It is possible to have

$$P(A|B) < P(A|B') \tag{3.45}$$

and have at the same time both

$$\begin{aligned} P(A|BC) &\geqq P(A|B'C), \\ P(A|BC') &\geqq P(A|B'C'). \end{aligned} \tag{3.46}$$

He offers the following hypothetical example:

| | $C$ | | | $C'$ | |
|---|---|---|---|---|---|
| | $B'$ | $B$ | | $B'$ | $B$ |
| $A'$ | 950 | 9000 | $A'$ | 5000 | 5 |
| $A$ | 50 (5%) | 1000 (10%) | $A$ | 5000 (50%) | 95 (95%) |

(3.47)

The marginal table, adding over the third variable in (3.47), is

| | $B'$ | $B$ |
|---|---|---|
| $A'$ | 5950 | 9005 |
| $A$ | 5050 (46%) | 1095 (11%) |

(3.48)

and the conditions in (3.45) and (3.46) are satisfied. But another way to interpret these conditions is that if we collapse a $2 \times 2 \times 2$ table over the third variable, the cross-product ratio for the marginal table ($\hat{\alpha} = 0.14$ in (3.48)) can be less than 1 while the cross product ratio in each layer (corresponding to levels of this third variable) is greater than 1 ($\hat{\alpha}_1 = 2.11$ and $\hat{\alpha}_2 = 19$ in (3.47)). Theorem 3.1 states that we can collapse over a variable and expect to find the same cross-product ratio in the margin as we have in each layer only if the variable collapsed over is independent of at least one of the other two variables. Thus we can view Simpson's paradox as a statement about the possible effects of collapsing multidimensional tables in the presence of interactions.

Both Meehl and Rosen [1955] and Lindley and Novick [1975] discuss Simpson's paradox, although from somewhat different perspectives. Darroch [1976] notes that it might be more appropriate to name the paradox after Yule, rather than Simpson, since Yule discussed it in the final section of his 1903 paper on the theory of association of attributes.

# 4
# Selection of a Model

## 4.1 General Issues

Complicated models involving large numbers of parameters most often fit a set of data more closely than a simpler model that is just a special case of the complicated one. (In the simplest contingency-table model considered in Chapter 3, the three-factor effect and the two-factor effects were set equal to zero, whereas the more complicated models allowed at least some nonzero two-factor effects.) On the other hand, a simple model is often to be preferred over a more complicated one that provides a better fit. There is thus a trade-off between goodness-of-fit and simplicity, and the dividing line between the "best" model and others that also fit the data adequately is clearly very fine (see the discussion of this point in Kruskal [1968]).

The goodness-of-fit tests discussed in Chapter 3 allow us to take a particular model and see if it yields expected values that are close in some sense to the observed values. For three-dimensional tables, there are eight possible hierarchical loglinear models (all containing the "main-effect" terms), any one of which might be used to fit the data. Because the test statistics are not statistically independent, we are not able to interpret the corresponding significance levels in the usual way. We simply cannot test the goodness-of-fit of each model separately, as we did in Chapter 3. Therefore we need a method that aids in the selection of the interaction terms to be included in the fitted model. Unfortunately, there is no all-purpose, best method of model selection. Bishop [1969], Brown [1976], Fienberg [1970b], Goodman [1970, 1971a], and Ku and Kullback [1968] all suggest different approaches to the model selection problem.

## 4.2 Conditional Test Statistics

Before discussing some approaches to model selection, it will be useful to introduce a general technique for comparing expected values for two different loglinear models, where one model is a special case of the second. Suppose that the two estimated expected values for the observed frequency $x_{ijk}$ are

$$\begin{aligned} \hat{m}_{ijk}(1) &= (\text{Expected})_1, \\ \hat{m}_{ijk}(2) &= (\text{Expected})_2, \end{aligned} \tag{4.1}$$

where model 2 is a special case of model 1 (e.g., the model of complete independence in a three-dimensional table is a special case of the model of independence of variable 3 with 1 and 2 jointly). Then the likelihood-ratio test statistic,

$$2\sum(\text{Observed}) \log \left[\frac{(\text{Expected})_1}{(\text{Expected})_2}\right], \tag{4.2}$$

can be used to test whether the difference between the expected values for the two models is simply due to random variation, given that the true expected values satisfy model 1. This conditional test statistic has an asymptotic chi-square distribution (under the null situation), with degrees of freedom equal to the difference in the degrees of freedom for the two models (for technical details see Rao [1973], pp. 418–420).

Goodman [1969] has noted that because of the multiplicative form of the estimated expected values for hierarchical loglinear models, expression (4.2) is equal to

$$2\sum(\text{Expected})_1 \log \left[\frac{(\text{Expected})_1}{(\text{Expected})_2}\right], \tag{4.3}$$

a test statistic analogous to the Pearson-like statistic proposed by Rao [1973, pp. 398–402]:

$$\sum \frac{[(\text{Expected})_1 - (\text{Expected})_2]^2}{(\text{Expected})_2}. \tag{4.4}$$

The methods discussed below are based mainly on expression (4.2).

We can use expression (4.2) to resolve our choice of models for the data in Table 3-6 (see Section 3.7 for the earlier discussion of this example). Since model (2) of Section 3.7, $u_{12} = u_{123} = 0$, is a special case of model (1), $u_{123} = 0$, the difference in the values of the likelihood-ratio statistics for the two models (from Table 3-7), $G^2(2) - G^2(1) = 50.9 - 25.1 = 25.8$, should be referred to a $\chi^2$ distribution table with $48 - 36 = 12$ d.f. to test if the true expected values satisfy model (2). This value corresponds to a descriptive level of significance somewhere between 0.05 and 0.01, and if we were to adhere to a 0.05 level of significance, we would choose model (1) over model (2).

## 4.3 Partitioning Chi-Square

The method of partitioning breaks the likelihood-ratio goodness-of-fit statistic for a hierarchical loglinear model into several additive parts. To make use of partitioning we must formulate a nested hierarchy of models in which we are interested, where each of the models considered must contain the previous ones in the hierarchy as special cases. For example, in a three-dimensional contingency table, the following sequence of models for the logarithms of the expected values forms a nested hierarchy:

$$\text{(a)}\quad u + u_{1(i)} + u_{2(j)} + u_{3(k)}, \tag{4.5}$$
$$\text{(b)}\quad u + u_{1(i)} + u_{2(j)} + u_{3(k)} + u_{13(ik)}, \tag{4.6}$$
$$\text{(c)}\quad u + u_{1(i)} + u_{2(j)} + u_{3(k)} + u_{13(ik)} + u_{23(jk)}, \tag{4.7}$$
$$\text{(d)}\quad u + u_{1(i)} + u_{2(j)} + u_{3(k)} + u_{12(ij)} + u_{13(ik)} + u_{23(jk)}. \tag{4.8}$$

A different nested hierarchy of models could be created by adding the two-factor effect terms in a different order. Depending on the hierarchy chosen, the partitioning method *may* yield different "best" models.

There are different strategies for constructing a hierarchy of models. If an investigator has auxiliary information regarding the interrelationships among the variables, he might set up the models so that the interaction terms added at successive stages are ordered according to their importance or expected magnitude. For three-dimensional tables one could, of course, look at all six possible hierarchies in order to determine if the choice of a hierarchy affects the choice of a model. The theorem on collapsing from Section 3.8, or the results given below in expressions (4.12) and (4.13), indicate that we can reduce this number to three hierarchies. Unfortunately, when we deal with higher-dimensional tables, the task of looking at all possible hierarchies is an onerous one, and the interpretation of the related tests of significance is difficult at best.

In the hierarchy given by (4.5)–(4.8) there are four possible models, and we use the partitioning technique for this hierarchy and the data from Table 3-2. We denote the likelihood-ratio goodness-of-fit statistics for models (4.5)–(4.8) by $G^2(\text{a})$, $G^2(\text{b})$, $G^2(\text{c})$, and $G^2(\text{d})$, respectively, and Table 3-5 indicates that $G^2(\text{a}) = 70.08$, $G^2(\text{b}) = 31.09$, $G^2(\text{c}) = 4.88$, and $G^2(\text{d}) = 3.02$. Note that

$$G^2(\text{a}) \geq G^2(\text{b}) \geq G^2(\text{c}) \geq G^2(\text{d}). \tag{4.9}$$

This result is true in general, and its importance will soon be evident. One of the reasons we do not partition the Pearson goodness-of-fit statistic, $X^2$, is that (4.9) does not necessarily hold for any set of nested models when $G^2$ is replaced by $X^2$. For example, in Table 3-5 (for the data from Table 3-1), we see that $X^2$ for model (4.5) is slightly less than $X^2$ for the model

$$u + u_{1(i)} + u_{2(j)} + u_{3(k)} + u_{12(ij)}. \tag{4.10}$$

Note that $G^2(\text{a}) - G^2(\text{b})$, $G^2(\text{b}) - G^2(\text{c})$, and $G^2(\text{c}) - G^2(\text{d})$ are statistics of the form (4.2) and can be used to test the difference between models (b) and (a), (c) and (b), and (d) and (c), respectively. Thus we can write the likelihood-ratio goodness-of-fit statistics for the model of complete independence as

$$G^2(\text{a}) = [G^2(\text{a}) - G^2(\text{b})] + [G^2(\text{b}) - G^2(\text{c})] + [G^2(\text{c}) - G^2(\text{d})] + G^2(\text{d}), \tag{4.11}$$

and each component has an asymptotic $\chi^2$ distribution with appropriate d.f. under a suitable null model.

The data in Table 3-2 indicate how a partitioning argument would go. We begin by looking at the component due to model (d), whose value 3.02 has a descriptive level of significance greater than 0.05 when referred to a table of the $\chi^2$ distribution with 1 d.f. This model fits the data fairly well, and we proceed to the next component, $G^2(\text{c}) - G^2(\text{d}) = 1.86$, whose descriptive level of significance is close to 0.20 when referred to the $\chi^2$ distribution with 1 d.f. Moreover, the cumulative value of the components so far, $G^2(\text{c}) = 4.88$, has a descriptive level of significance slightly less than 0.10 when referred to the $\chi^2$ distribution with 2 d.f. Thus model (4.7) seems to fit well, and we go on to the third component, $G^2(\text{b}) - G^2(\text{c}) = 26.21$, which exceeds the 0.001 upper tail value of the $\chi^2$ distribution with 1 d.f. by a considerable amount. It thus makes sense to stop the partitioning here and choose model (c), conditional independence of perch height and perch diameter given species. Were we to continue examining components we would find that $G^2(\text{a}) - G^2(\text{b}) = 38.99$ is also highly significant, but this will not necessarily be the case for partitioning analyses in other examples.

At each stage of the partitioning we should look at two things: the value of the appropriate component, and the cumulative value of the components examined, which is equivalent to the $G^2$ statistic for the simple model examined so far. One possible rule for the partitioning technique is: Stop partitioning when either of these values is significant at some level, when referred to the appropriate $\chi^2$ distribution, and choose as the best model the one from the previous step.

Choosing the appropriate hierarchy of models is very important. To illustrate this point, consider an alternate hierarchy of models consisting of

$$\begin{aligned}
&(\text{a}') \quad u + u_{1(i)} + u_{2(j)} + u_{3(k)},\\
&(\text{b}') \quad u + u_{1(i)} + u_{2(j)} + u_{3(k)} + u_{12(ij)},\\
&(\text{c}') \quad u + u_{1(i)} + u_{2(j)} + u_{3(k)} + u_{12(ij)} + u_{13(ik)},\\
&(\text{d}') \quad u + u_{1(i)} + u_{2(j)} + u_{3(k)} + u_{12(ij)} + u_{13(ik)} + u_{23(jk)}.
\end{aligned}$$

From Table 3-5 (for the data in Table 3-2), we have $G^2(\text{d}') = G^2(\text{d}) = 3.02$ as before, but now the remaining three components,

$$G^2(\text{c}') - G^2(\text{d}') = 15.38,$$

$$G^2(\mathrm{b}') - G^2(\mathrm{c}') = 38.99,$$
$$G^2(\mathrm{a}') - G^2(\mathrm{b}') = 12.69,$$

are all greater than the 0.001 upper tail value of the $\chi^2$ distribution with 1 d.f. Thus, based on this partition of $G^2(\mathrm{a}')$, we are forced to settle on (d′), no three-factor interaction, as the most appropriate model for the data.

The method of partitioning was first introduced by Lancaster [1951] for the Pearson test statistic $X^2$ (see also Lancaster [1969]) and was later extended by Kullback, Kupperman, and Ku [1962] and Ku and Kullback [1968] for the likelihood-ratio statistic $G^2$. Unfortunately, the methods of Lancaster, Kullback, Kupperman, and Ku are somewhat different from those described here, and they cannot be used as a substitute for the procedures in this section. In particular, the method of Kullback, Kupperman, and Ku [1962] can yield negative values for the likelihood-ratio goodness-of-fit statistic in tests for three-way or higher interactions. This is due to an improper procedure used by these authors to estimate expected cell counts under the no three-factor interaction model. The correct procedure is now known and is the one described here. Thus it is *not* possible to get negative values for the likelihood-ratio statistics in the partitioning method described here, except as a result of numerical roundoff errors associated with the iterative proportional fitting procedure, and such errors tend to be quite small. Darroch [1974, 1976] gives further details on the Lancaster method and the models for which it is suitable.

Goodman [1969, 1970] describes an elaborate method for partitioning the likelihood-ratio goodness-of-fit statistics. He not only carries out the partitioning as described above, but he also gives an additional interpretation to several of the components due to differences between models, and he offers a general set of rules for this more elaborate method of partitioning. These rules have been elaborated upon by Bishop, Fienberg, and Holland [1975]. An interesting special case of Goodman's results is that certain conditional tests are identical to corresponding marginal tests. For example, in the hierarchy of models given by (4.5)–(4.7), the conditional test statistic for model (b) given model (c) is algebraically identical to the marginal test statistic for independence in the two-dimensional table for variables 2 and 3; that is,

$$G^2(\mathrm{b}|\mathrm{c}) = 2\sum_{i,j,k} x_{ijk} \log\left[\frac{x_{i+k}x_{+jk}/x_{++k}}{x_{+j+}x_{i+k}/N}\right]$$
$$= 2\sum_{i,j,k} x_{ijk} \log\left[\frac{x_{+jk}}{x_{+j+}x_{++k}/N}\right] \tag{4.12}$$

$$= 2 \sum_{j,k} x_{+jk} \log \left[ \frac{x_{+jk}}{x_{+j+}x_{++k}/N} \right].$$

Similarly, the conditional test statistic for model (a) given model (b) is algebraically identical to the marginal test statistic for independence between variables 1 and 3; that is,

$$\begin{aligned} G^2(\mathrm{a}|\mathrm{b}) &= 2 \sum_{i,j,k} x_{ijk} \log \left[ \frac{x_{+j+}x_{i+k}/N}{x_{i++}x_{+j+}x_{++k}/N^2} \right] \\ &= 2 \sum_{i,j,k} x_{ijk} \log \left[ \frac{x_{i+k}}{x_{i++}x_{++k}/N} \right] \\ &= 2 \sum_{i,k} x_{i+k} \log \left[ \frac{x_{i+k}}{x_{i++}x_{++k}/N} \right]. \end{aligned} \tag{4.13}$$

Thus the use of the two-dimensional marginal test statistic to test for $u_{12} = 0$ is justified provided $u_{23} = 0$, $u_{13} = 0$, or both. This is a byproduct of the somewhat more general theorem on collapsing discussed in Section 3.8.

Goodman [1971a] suggests the use of stepwise procedures for model selection, by analogy with the stepwise methods of multiple regression. These methods are mostly of interest for tables with four or more dimensions, so we defer our discussion of them until the next chapter.

## 4.4 Using Information about Ordered Categories

One characteristic of the techniques used up to this point is that they have made no assumptions regarding the ordering of categories for any of the underlying variables. In many situations, however, the categories for a variable represent points in an ordinal scale; in political preference polls, for example, respondents are often asked to rate the incumbent's position according to a five-point scale: agree strongly, agree, indifferent, disagree, disagree strongly. If this information on the ordered structure of the categories is ignored, all the relevant aspects of the data have not been utilized.

We have not completely ignored information regarding ordered categories in earlier analyses. For example, after fitting a reasonable loglinear model to the NBER–TH data in Section 3.7, we looked at the values of the estimated interaction $u$-terms and interpreted them in light of the underlying ordering of the categories. But this was done informally, and the discussion might have been strengthened by a more structured approach to the problem.

A number of authors have proposed methods for dealing with ordered categories in multidimensional contingency tables (e.g., Bock [1975], Clayton [1974], Grizzle, Starmer, and Koch [1969], and Williams and Grizzle [1972]). The approach described here is essentially the same as those presented in Haberman [1974b], Nerlove and Press [1973], and Simon [1974].

For simplicity we begin with an $I \times J$ two-dimensional contingency table with observed entries $\{x_{ij}\}$ generated by one of the three basic sampling models. Suppose that the $J$ columns are ordered and that a priori we can assign scores $\{v_j\}$ to them. If we decide that the model of independence does not hold, rather than going directly to the general saturated loglinear model it might be logical to explore an interaction structure that directly reflects the ordering of the columns and of the scores $\{v_j\}$, such as

$$\log m_{ij} = u + u_{1(i)} + u_{2(j)} + (v_j - \bar{v})u'_{1(i)}, \tag{4.14}$$

where $\bar{v}$ is the average of the $\{v_j\}$ and the usual side constraints hold:

$$\sum_i u_{1(i)} = \sum_j u_{2(j)} = \sum_i u'_{1(i)} = 0. \tag{4.15}$$

Note that the $\{v_j\}$ have been centered in the interaction term so that

$$\sum_j u_{12(ij)} = \sum_j (v_j - \bar{v})u'_{1(i)} = 0, \tag{4.16}$$

in keeping with the convention used elsewhere in this book. This normalization is really not necessary in the present circumstances.

The likelihood equations for model (4.14) follow the general format of the likelihood equations for other loglinear models discussed in this book:

$$\hat{m}_{i+} = x_{i+}, \qquad i = 1, 2, \ldots, I, \tag{4.17}$$

$$\hat{m}_{+j} = x_{+j}, \qquad j = 1, 2, \ldots, J, \tag{4.18}$$

and

$$\sum_j v_j \hat{m}_{ij} = \sum_j v_j x_{ij}, \qquad i = 1, 2, \ldots, I. \tag{4.19}$$

The symmetry inherent in the sets of equations (4.17) and (4.18) is important. To put (4.19) into a similar form we transform the $\{v_j\}$ to $\{v_j^*\}$, where $0 \le v_j^* \le 1$ for $j = 1, 2, \ldots J$. Then (4.19) becomes

$$\sum_j v_j^* \hat{m}_{ij} = \sum_j v_j^* x_{ij}, \qquad i = 1, 2, \ldots, I, \tag{4.20}$$

and by subtracting (4.20) from (4.17) we also have

$$\sum_j (1 - v_j^*)\hat{m}_{ij} = \sum_j (1 - v_j^*)x_{ij}, \qquad i = 1, 2, \ldots, I. \tag{4.21}$$

Equation (4.21) is redundant once we have (4.20), in the same sense that $\hat{m}_{i+} = x_{i+}$ for $i = 1, 2, \ldots, I - 1$ implies $\hat{m}_{I+} = x_{I+}$.

The likelihood equations (4.17), (4.18), (4.20), and (4.21) can be solved by a version of iterative proportional fitting in which we set

$$\hat{m}_{ij}^{(0)} = 1, \qquad i = 1, 2, \ldots, I, \quad j = 1, 2, \ldots, J \tag{4.22}$$

and for $v \geq 0$ cycle through the three steps:

$$\hat{m}_{ij}^{(3v+1)} = \hat{m}_{ij}^{(3v)}\left(\frac{x_{i+}}{\hat{m}_{i+}^{(3v)}}\right), \tag{4.23}$$

$$\hat{m}_{ij}^{(3v+2)} = \hat{m}_{ij}^{(3v+1)}\left(\frac{x_{+j}}{\hat{m}_{+j}^{(3v+1)}}\right), \tag{4.24}$$

$$\hat{m}_{ij}^{(3v+3)} = \hat{m}_{ij}^{(3v+2)}\left(\frac{\Sigma_k v_k^* x_{ik}}{\Sigma_k v_k^* \hat{m}_{ik}^{(3v+2)}}\right)^{v_j^*}\left(\frac{\Sigma_k(1 - v_k^*)x_{ik}}{\Sigma_k(1 - v_k^*)\hat{m}_{ik}^{(3v+2)}}\right)^{1 - v_j^*}, \tag{4.25}$$

for $i = 1, 2, \ldots, I$ and $j = 1, 2, \ldots, J$. Repeated application of (4.23)–(4.25) for $v = 0, 1, 2, \ldots$ gives an iterative procedure that converges to the estimated expected values $\{\hat{m}_{ij}\}$ (for a proof of convergence see Darroch and Ratcliff [1972]). Note that the adjustment for equations (4.20) and (4.21) is done in (4.25) simultaneously, and that the two multiplicative adjustment factors associated are raised to the powers $v_j^*$ and $1 - v_j^*$, respectively.

While the version of iterative proportional fitting used to solve the likelihood equations in this problem does converge, the rate of convergence in practice is considerably slower than the rate for the standard iterative proportional fitting procedure. Haberman [1974b] describes a Newton–Raphson algorithm to handle this problem that converges more rapidly than the algorithm outlined above. The speed of convergence of the Newton–Raphson algorithm must be weighed against the simplicity of the structure of equations (4.23)–(4.25).

Since we are estimating $I - 1$ additional parameters relative to the model of independence, the degrees of freedom associated with (4.14) equal $(I - 1)(J - 1) - (I - 1) = (I - 1)(J - 2)$. Clearly this approach only makes sense if $J > 2$. The goodness-of-fit of this model may be tested with either

**Table 4-1**
Frequency of Visits by Length of Stay for 132 Long-Term Schizophrenic Patients (Wing [1962])

| Frequency of visiting | Length of stay in hospital: At least 2 years but less than 10 years | At least 10 years but less than 20 years | At least 20 years | Totals |
|---|---|---|---|---|
| Goes home, or visited regularly | 43 | 16 | 3 | 62 |
| Visited less than once a month. Does not go home. | 6 | 11 | 10 | 27 |
| Never visited and never goes home | 9 | 18 | 16 | 43 |
| Totals | 58 | 45 | 29 | 132 |

$X^2$ or $G^2$ in the usual fashion.

Haberman [1974b] presents an analysis of data given by Wing [1962] comparing frequency of visits with length of stay for 132 long-term schizophrenic patients in two London mental hospitals (Table 4-1). Fitting the model of independence to these data yields $X^2 = 35.2$ and $G^2 = 38.4$ with 4 d.f. Fitting the "ordered-interaction" model (4.14), with $v_1 = -1$, $v_2 = 0$, $v_3 = 1$, and the added constraint that $u_{2(j)} = (v_j - \bar{v})u^*$, yields $X^2 = 3.3$ and $G^2 = 3.2$ with 3 d.f. Using the ordered restrictions and the extra degree of freedom in the example provides a remarkable improvement in fit. Expected values for this model are in Table 4-2.

The approach just described generalizes immediately to handle (1) tables of dimension three or more, (2) ordered categories for more than one variable, and (3) quadratic and higher-order components, in a manner analogous to the handling of such problems in the analysis of variance. The only thing about which we need to be careful is that once quadratic and higher-order components are set equal to zero for a particular set of $u$-terms, they are also set equal to zero for all related higher-order $u$-terms. This restriction preserves the hierarchical structure of the loglinear models being used. For example, suppose we have an $I \times J \times K$ table (where $J > 2$) with ordered columns having assigned scores $\{v_j\}$, and we wish to explore the linearity of various interaction effects involving the columns. If we set

$$u_{12(ij)} = (v_j - \bar{v})u'_{1(i)} \tag{4.26}$$

**Table 4-2**
MLEs of Expected Values under Model (4.14) for Data in Table 4-1

| Frequency of visiting | Length of stay in hospital | | |
|---|---|---|---|
| | At least 2 years but less than 10 years | At least 10 years but less than 20 years | At least 20 years |
| Goes home, or visited regularly | 44.19 | 13.61 | 4.19 |
| Visited less than once a month. Does not go home. | 7.07 | 8.85 | 11.07 |
| Never visited and never goes home | 10.98 | 14.05 | 17.98 |

in the general loglinear model, as we did in (4.14), then we must also set

$$u_{123(ijk)} = (v_j - \bar{v})u'_{13(ik)}, \tag{4.27}$$

where

$$\sum_i u'_{13(ik)} = \sum_k u'_{13(ik)} = 0. \tag{4.28}$$

Forcing expression (4.27) to follow from (4.26) is similar to the restrictions associated with hierarchical models (e.g., $u_{12} = 0$ implies $u_{123} = 0$). Such restrictions are needed to simplify the interpretation of our models and also for technical reasons not discussed here.

The degrees of freedom for this model equal $(I - 1)(J - 2)K$. Following the general results for loglinear models in Appendix II, we can write the likelihood equations for this model as

$$\hat{m}_{i+k} = x_{i+k}, \qquad i = 1, 2, \ldots, I, \quad k = 1, 2, \ldots, K, \tag{4.29}$$

$$\hat{m}_{+jk} = x_{+jk}, \qquad j = 1, 2, \ldots, J, \quad k = 1, 2, \ldots, K, \tag{4.30}$$

$$\sum_j v_j \hat{m}_{ijk} = \sum_j v_j x_{ijk}, \qquad i = 1, 2, \ldots, I, \quad k = 1, 2, \ldots, K. \tag{4.31}$$

If the model is specified only by (4.27), then the likelihood equations consist of

(4.29)–(4.31), and

$$\hat{m}_{ij+} = \hat{x}_{ij+}, \qquad i = 1, 2, \ldots, I, \quad j = 1, 2, \ldots, J, \tag{4.32}$$

whereas for the model given by (4.26) and $u_{123(ijk)} = 0$, the likelihood equations consist of (4.29), (4.30), and

$$\sum_j v_j \hat{m}_{ij+} = \sum_j v_j x_{ij+}, \qquad i = 1, 2, \ldots, I. \tag{4.33}$$

For any of these models, we can solve the likelihood equations using a version of iterative proportional fitting in which each cycle of the iteration has a step to adjust for one of the sets of likelihood equations. Alternatively, Haberman's [1974b] Newton–Raphson algorithm is especially useful when we wish to explore quadratic and higher-order components of the interaction involving variables with ordered categories.

Ashford and Sowden [1970] present data on British coal miners between the ages of 20 and 64 years who were smokers but did not show radiological signs of pneumoconiosis. The data, consisting of information on two symptoms (breathlessness and wheeze) and age, are given in Table 4-3. Age is represented in terms of nine 5-year groupings. The analysis here is similar to that given by Plackett [1974].

A cursory examination of Table 4-3 suggests that the relationship between breathlessness (variable 1) and wheezing (variable 2) decreases with age

**Table 4-3**
Coal Miners Classified by Age, Breathlessness, and Wheeze (Ashford and Sowden [1970])

| Age group | Breathlessness | | No breathlessness | | |
|---|---|---|---|---|---|
| in years | Wheeze | No wheeze | Wheeze | No wheeze | Totals |
| 20–24 | 9 | 7 | 95 | 1841 | 1952 |
| 25–29 | 23 | 9 | 105 | 1654 | 1791 |
| 30–34 | 54 | 19 | 177 | 1863 | 2113 |
| 35–39 | 121 | 48 | 257 | 2357 | 2783 |
| 40–44 | 169 | 54 | 273 | 1778 | 2274 |
| 45–49 | 269 | 88 | 324 | 1712 | 2393 |
| 50–54 | 404 | 117 | 245 | 1324 | 2090 |
| 55–59 | 406 | 152 | 225 | 967 | 1750 |
| 60–64 | 372 | 106 | 132 | 526 | 1136 |
| Totals | 1827 | 600 | 1833 | 14,022 | 18,282 |

**Table 4-4**
Estimated Expected Values under Model of Expression (4.27) for Ashford–Sowden Data in Table 4-3

| Age group in years | Breathlessness Wheeze | Breathlessness No wheeze | No breathlessness Wheeze | No breathlessness No wheeze | Totals |
|---|---|---|---|---|---|
| 20–24 | 10.2 | 5.8 | 93.8 | 1842.2 | 1952 |
| 25–29 | 21.2 | 10.8 | 106.8 | 1652.2 | 1791 |
| 30–34 | 52.5 | 20.5 | 178.5 | 1861.5 | 2113 |
| 35–39 | 121.4 | 47.6 | 256.6 | 2357.0 | 2783 |
| 40–44 | 169.3 | 53.7 | 272.7 | 1778.3 | 2274 |
| 45–49 | 274.8 | 82.2 | 318.2 | 1717.8 | 2393 |
| 50–54 | 393.3 | 127.7 | 255.7 | 1313.3 | 2090 |
| 55–59 | 418.8 | 139.2 | 212.2 | 979.8 | 1750 |
| 60–64 | 365.5 | 112.5 | 138.5 | 519.5 | 1136 |
| Totals | 1827 | 600 | 1833 | 14,022 | 18,282 |

(variable 3). For the loglinear model that posits no second-order interaction ($u_{123} = 0$), the likelihood-ratio statistic is $G^2 = 26.7$ with 8 d.f., a highly significant value. The linear second-order interaction model of expression (4.27), with $v_j = j$, yields $G^2 = 6.8$ with 7 d.f., an extremely good fit. The expected values under this model are listed in Table 4-4, and the estimated value of the one parameter used to describe the second-order interaction is $\hat{u}'_{13(11)} = -0.13$.

We shall return to the discussion of ordered effects associated with ordered categories in Chapter 6, where we consider logistic response models.

# 5
# Four- and Higher-Dimensional Contingency Tables

Although all the models and methods discussed so far have been in the context of three-dimensional tables, the extensions to higher-dimensional tables are relatively straightforward. To illustrate these extensions to fourway tables we shall use the data in Table 5-1, which were studied earlier by Ries and Smith [1963], Cox and Lauh [1967], Goodman [1971a], Bishop, Fienberg, and Holland [1975], and others. The data come from an experiment in which a sample of 1008 individuals was asked to compare two detergents, a new product X and a standard product M, placed with members of the sample. In addition to assessing brand preferences (X or M) the experimenters inquired about whether sample members had used brand M previously (yes or no), about the degree of softness of the water they used (soft, medium, hard), and about the temperature of the laundry water used (high, low). We refer to "softness," "use," "temperature," and "preference" as variables 1, 2, 3, and 4, respectively. The table is of dimension $3 \times 2 \times 2 \times 2$, and we note that preference is a response variable and the other three are explanatory variables. While we shall not make use of this information in our present analysis of these data, we shall take advantage of the response–explanatory distinction in a reanalysis of the data in Section 6.2.

## 5.1 The Loglinear Models and MLEs for Expected Values

Suppose for a four-dimensional $I \times J \times K \times L$ table that the total of the counts is $N$ and that the count for the $(i,j,k,l)$ cell is $x_{ijkl}$. As usual, when the cell frequencies are added over a particular variable, we replace the subscript

**Table 5-1**
Cross-Classification of Sample of 1008 Consumers According to (1) the Softness of the Laundry Water Used, (2) the Previous Use of Detergent Brand M, (3) the Temperature of the Laundry Water Used, (4) the Preference for Detergent Brand X over Brand M in a Consumer Blind Trial (Ries and Smith [1963])

| Water softness | Brand preference | Previous user of M High temperature | Previous user of M Low temperature | Previous nonuser of M High temperature | Previous nonuser of M Low temperature |
|---|---|---|---|---|---|
| Soft | X | 19 | 57 | 29 | 63 |
| | M | 29 | 49 | 27 | 53 |
| Medium | X | 23 | 47 | 33 | 66 |
| | M | 47 | 55 | 23 | 50 |
| Hard | X | 24 | 37 | 42 | 68 |
| | M | 43 | 52 | 30 | 42 |

for that variable by a "+". We denote by $m_{ijkl}$ the expected cell value for the $(i, j, k, l)$ cell under some parametric model, and we use the same summation notation for the expected values.

The simplest model for a four-dimensional table corresponds to the complete independence of all four variables, and for this model we can write the natural logarithm of the expected cell frequencies in the form

$$\log m_{ijkl} = u + u_{1(i)} + u_{2(j)} + u_{3(k)} + u_{4(l)}, \tag{5.1}$$

with the usual ANOVA-like constraints:

$$\sum_i u_{1(i)} = \sum_j u_{2(j)} = \sum_k u_{3(k)} = \sum_l u_{4(l)} = 0. \tag{5.2}$$

In our abbreviated notation we refer to this model as [1] [2] [3] [4]. Since it is highly unlikely that the four variables for the data in Table 5-1 are completely independent, we require more complex loglinear models that include two-factor and higher-order interaction terms. There are $\binom{4}{2} = 6$ possible sets of two-factor terms such as $\{u_{12(ij)}\}$, $\binom{4}{3} = 4$ possible sets of three-factor terms such as $\{u_{123(ijk)}\}$, and 1 set of four-factor terms $\{u_{1234(ijkl)}\}$. We restrict ourselves to the use of hierarchical loglinear models in which, whenever we include an interaction term, we must also include all lower-order interactions involving variables in the higher-order term. Thus, if we include $u_{123(ijk)}$ in a loglinear model, we must also include $u_{12(ij)}$, $u_{13(ik)}$, and $u_{23(jk)}$. Conversely, if $u_{12(ij)} = 0$ for all values of $i$ and $j$, then

$$u_{123(ijk)} = u_{124(ijl)} = u_{1234(ijkl)} = 0 \tag{5.3}$$

for all $i$, $j$, $k$, and $l$. The restriction to hierarchical models allows us to use a simple generalization of the iterative proportional fitting procedure, described in Chapter 3, for computing MLEs of the expected cell values.

For four-dimensional tables there are 113 different hierarchical loglinear models, all of which include the main-effect $u$-terms in expression (5.1). Good [1975] has addressed the problem of enumerating the possible models of independence (both mutual independence and conditional independence) in an $m$-dimensional table. The number of such models grows extremely rapidly with $m$, and for $m = 10$ there are 3,475,978. This number is considerably smaller than the total number of different hierarchical loglinear models.

As in Chapter 3, there usually exist unique nonzero MLEs for the expected cell counts under various hierarchical loglinear models when the observed data are generated by any one of the following three sampling schemes:
(1) a Poisson sampling scheme, with independent Poisson random variables

for each cell (usually based on observations from a set of Poisson processes for a given period of time);

(2) a single, common multinomial sampling scheme;

(3) a set of multinomial sampling schemes, each multinomial corresponding to one entry in a given set of fixed marginal totals.

In scheme (1) nothing is fixed by the sampling model; in scheme (2) the total number of observations $N$ is fixed; and in scheme (3) one set of marginal totals is fixed. These three sampling schemes lead to the same MLEs if we include in our model the $u$-terms corresponding to the marginal totals fixed in scheme (3). The detergent example we are considering here resulted from sampling scheme (2), although in Chapter 6 we shall analyze the data using a conditional argument that corresponds to sampling scheme (3).

For many sets of data we can usually get by with models that include only two-factor interaction terms. (This is fortunate because it is easier to interpret two-factor interactions than it is to interpret those involving three or four factors.) Nevertheless, we still must use an iterative computing procedure to get the MLEs of the expected cell values for most loglinear models. Two exceptions, where we can write down MLEs directly, are the complete-independence model, given by expression (5.1), for which we get

$$\hat{m}_{ijkl} = \frac{x_{i+++}x_{+j++}x_{++k+}x_{+++l}}{N^3}, \tag{5.4}$$

and the conditional-independence model with $u_{12(ij)} = 0$ for all $i$ and $j$ (by our restriction to hierarchical models, this implies that $u_{123}$, $u_{124}$, and $u_{1234}$ are also zero), for which we get

$$\hat{m}_{ijkl} = \frac{x_{i+kl}x_{+jkl}}{x_{++kl}}. \tag{5.5}$$

Expressions (5.4) and (5.5) are the natural generalizations of (3.16) and (3.22). Bishop [1971], Goodman [1971b], Bishop, Fienberg, and Holland [1975], and Sundberg [1975] give rules for deciding whether or not the MLEs for a particular model can be found directly; however, the method of iterative proportional fitting can always be used to compute the MLEs.

We now consider the model with all two-factor terms but no higher-order terms:

$$\begin{aligned} u &+ u_{1(i)} + u_{2(j)} + u_{3(k)} + u_{4(l)} \\ &+ u_{12(ij)} + u_{13(ik)} + u_{14(il)} \\ &+ u_{23(jk)} + u_{24(jl)} + u_{34(kl)}. \end{aligned} \tag{5.6}$$

In our abbreviated notation this model is [12] [13] [14] [23] [24] [34], or "all two-way." Applying the rules of Bishop and Goodman to this model we find that a generalization of the iterative proportional fitting procedure must be used to get estimated expected cell values. To carry out the computations we need the observed marginal totals corresponding to the highest-order interactions involving the variables; that is, we need the six different sets of two-dimensional marginal totals corresponding to the two-factor effects. Each cycle of the iteration has six different steps corresponding to the adjustment for the six sets of marginal totals.

The notions of testing for goodness-of-fit generalize immediately to cover this type of multidimensional table. Degrees of freedom to be associated with various goodness-of-fit test statistics are determined in a manner analogous to that for three-dimensional tables. For example, when there are $I$, $J$, $K$ and $L$ levels for variables, 1, 2, 3, and 4, respectively, the degrees of freedom for model (5.6) are equal to

$$\begin{aligned} IJKL - [&1 + (I-1) + (J-1) + (K-1) + (L-1) + (I-1)(J-1) \\ &+ (I-1)(K-1) + (I-1)(L-1) + (J-1)(K-1) \\ &+ (J-1)(L-1) + (K-1)(L-1)] \end{aligned} \tag{5.7}$$

(see also the listing for this model in Table 5-2). In the case of Table 5-1, where $I = 3$, $J = 2$, $K = 2$, and $L = 2$, expression (5.7) reduces to 9 d.f.

In Table 5-2 are listed examples of various loglinear models, in the abbreviated notation, and for each the d.f. associated with the model in an $I \times J \times K \times L$ table is given, along with the actual values for the case $I = J = K = L = 2$. Following Bishop [1971] and Goodman [1970], the models are separated into two groups, those for which the MLEs can be found directly and those for which iteration is required.

Table 5-3 gives values of the likelihood-ratio chi-square statistic, $G^2$, and the associated degrees of freedom for eleven different loglinear models (denoted by the abbreviated notation) fitted to the Ries–Smith data of Table 5-1. We shall make use of much of this information in the discussion of model selection in the next two sections.

Given the large number of possible hierarchical loglinear models that can be fit to a multidimensional table, it is reasonable to ask whether a relatively systematic approach to model selection is possible. Many different approaches have been proposed, none of them entirely satisfactory. Some of these approaches are described in the remainder of this chapter.

### 5.2 Using Partitioning to Select a Model

Just as in Chapter 4, we can use the partitioning of the likelihood-ratio chi-

**Table 5-2**
Degrees of Freedom Associated with Various Loglinear Models for Four-Dimensional Tables

| Model Abbreviation | d.f.* |
|---|---|
| **I. Models with Direct Estimates** | |
| (a) [1][2][3][4] | 11<br>$[IJKL - I - J - K - L + 3]$ |
| (b) [12][3][4] | 10<br>$[IJKL - IJ - K - L + 2]$ |
| (c) [12][34] | 9<br>$[(IJ - 1)(KL - 1)]$ |
| (d) [12][23][4] | 9<br>$[J(IKL - I - K + 1) - L + 1]$ |
| (e) [12][23][34] | 8<br>$[IJK - IJ - KL - JK + J + K]$ |
| (f) [12][13][14] | 8<br>$[I(JKL - J - K - L + 2)]$ |
| (g) [123][4] | 7<br>$[(IJK - 1)(L - 1)]$ |
| (h) [123][34] | 6<br>$[K(IJ - 1)(L - 1)]$ |
| (i) [123][234] | 4<br>$[JK(I - 1)(L - 1)]$ |
| **II. Models Requiring Indirect Estimates** | |
| (j) [12][13][23][4] | 8<br>$[IJKL - IJ - JK - IK - L + I + J + K]$ |
| (k) [12][13][23][34] | 7<br>$[IJKL - IJ - JK - IK - KL + I + J + 2K - 1]$ |
| (l) [12][13][23][24][34] | 6<br>$[IJKL - IJ - JK - IK - JL - KL + I + 2J + 2K + L - 2]$ |
| (m) all two-way | 5<br>$[IJKL - IJ - IK - IL - JK - JL - KL + 2(I + J + K + L) - 3]$ |
| (n) [123][24][34] | 5<br>$[IJKL - IJK - JL - KL + J + K + L - 1]$ |
| (o) [123][14][24][34] | 4<br>$[IJKL - IJK - IL - JL - KL + I + J + K + 2L - 2]$ |
| (p) [123][124][34] | 3<br>$[(IJ - 1)(K - 1)(L - 1)]$ |
| (q) [123][124][234] | 2<br>$[(IJ - J + 1)(K - 1)(L - 1)]$ |
| (r) all three-way | 1<br>$[(I - 1)(J - 1)(K - 1)(L - 1)]$ |

*The first entry pertains to the $2 \times 2 \times 2 \times 2$ table and the second entry to the $I \times J \times K \times L$ table.

**Table 5-3**
Likelihood Ratio Chi-Square Values for Some Loglinear Models Applied to the Data in Table 5-1

| Model | d.f. | $G^2$ |
|---|---|---|
| [1] [2] [3] [4] | 18 | 42.9 |
| [12] [13] [14] [23] [24] [34] | 9 | 9.9 |
| [123] [124] [134] [234] | 2 | 0.7 |
| [1] [3] [24] | 17 | 22.4 |
| [1] [24] [34] | 16 | 18.0 |
| [13] [24] [34] | 14 | 11.9 |
| [13] [23] [24] [34] | 13 | 11.2 |
| [12] [13] [23] [24] [34] | 11 | 10.1 |
| [1] [234] | 14 | 14.5 |
| [134] [24] | 10 | 12.2 |
| [13] [234] | 10 | 8.4 |
| [24] [34] [123] | 9 | 8.4 |
| [123] [234] | 8 | 5.6 |

square statistic to select a loglinear model that describes a four-dimensional table of counts. We illustrate this approach on the Ries–Smith data given in Table 5-1, where "softness," "use," "temperature," and "preference" are variables 1, 2, 3, and 4, respectively.

The simplest model we consider has no interactions; in the abbreviated notation it is

(a) [1] [2] [3] [4].

It would be natural for brand preference to be related to use, so our next model includes this interaction:

(b) [24] [1] [3].

Since some detergents are designed for cold water and some for hot water, it would not be unreasonable for preference to be related to temperature as well:

(c) [24] [34] [1].

Next, we add the interaction between softness and temperature, since the softer the water, the higher the temperature the detergent manufacturers usually suggest:

(d) [13] [24] [34].

Finally, we include two models that incorporate three-factor interactions:

(e) [13] [234],

(f) [123] [234].

Table 5-4 gives the results of partitioning the likelihood-ratio chi-square statistic for model (a) as applied to these data, using the hierarchy of six models just constructed and the $G^2$ values from Table 5-3. Starting at the bottom of the table and working our way up, we find that the first component to exceed the upper 0.05 tail value of the corresponding $\chi^2$ distribution is the one due to the difference between models (d) and (c). This suggests that model (d) is the "best" model. We note that the values of $G^2$ for models (c) and (b) are still not significant at the 0.05 level. Moreover, the differences between models (d) and (c) and between (c) and (b) are not significant at the 0.01 level. Thus we might still wish to consider models (b) and (c) as possible competitors to model (d) for "best" model.

## 5.3 Stepwise Selection Procedures

We now present a brief description of the stepwise procedures for model selection suggested by Goodman [1971a]. The description here is for four-dimensional tables, but can easily be extended to tables with five or more dimensions. A similar approach would be appropriate for a linear model in a fully crossed layout with unequal numbers of observations in the cells and a continuous response variate.

We begin by choosing a significance level, say 0.05, and then we test for

**Table 5-4**
A Partitioning of the Likelihood Ratio Chi-Square Statistic for Complete Independence as Applied to the Ries–Smith Data in Table 5-1

| Component due to | $G^2$ | d.f. |
|---|---|---|
| Model (a) | 42.9* | 18 |
| Difference between models (b) and (a) | 20.5* | 1 |
| Model (b) | 22.4 | 17 |
| Difference between models (c) and (b) | 4.4* | 1 |
| Model (c) | 18.0 | 16 |
| Difference between models (d) and (c) | 6.1* | 2 |
| Model (d) | 11.9 | 14 |
| Difference between models (e) and (d) | 3.5 | 4 |
| Model (e) | 8.4 | 10 |
| Difference between models (f) and (e) | 2.8 | 2 |
| Model (f) | 5.6 | 8 |

* Indicates that value is in upper 5% tail of the corresponding $\chi^2$ distribution, with d.f. as indicated.

the goodness-of-fit of the three models

(1) $u_{12} = u_{13} = u_{14} = u_{23} = u_{24} = u_{34} = 0$,
(2) $u_{123} = u_{124} = u_{134} = u_{234} = 0$,
(3) $u_{1234} = 0$.

(Note that (1) implies (2) and (2) implies (3).) If model (3) does not fit the data, we stop and choose as our model the general one with all $u$-terms present. (It is possible to have model (1) or (2) fit, while model (3) does not. Such a situation occurs rarely, and Goodman does not discuss it.) If model (3) fits but model (2) does not, we choose:

(A) *for forward selection:* model (2), and we add three-factor $u$-terms (as described below).

(B) *for backward elimination:* model (3), and we delete three-factor $u$-terms.

If models (2) and (3) fit but (1) does not, we choose:

(C) *for forward selection:* model (1), and we add two-factor $u$-terms.

(D) *for backward elimination:* model (2), and we delete two-factor $u$-terms.

Suppose models (2) and (3) fit but (1) does not. Then the steps involved in forward sclcction arc:

(C1) Add that two-factor $u$-term whose conditional goodness-of-fit statistic of the form (4.2) is most significant, provided the descriptive level of significance (i.e., the $p$-value) does not exceed the preselected value, say 0.05.

(C2) Add the next most significant two-factor $u$-term, using the conditional test statistic involving the model from the preceding step.

(C3) (Optional) Delete any two-factor $u$-terms that no longer make a significant contribution to the model (using appropriate conditional test statistics).

(C4) Repeat steps C2 and C3 until no further two-factor terms can be added or dropped.

The steps involved in backward elimination are similar but go in the reverse direction:

(D1) Eliminate the least significant two-factor $u$-term, given that the descriptive level of significance exceeds the preselected value, say 0.05.

(D2) Eliminate the next least significant two-factor $u$-term, using the conditional test statistic involving the model from the preceding step.

(D3) Add back two-factor terms that now significantly improve the fit of the model.

(D4) Repeat steps D2 and D3 until no further two-factor terms can be added or deleted.

We illustrate these selection procedures on the Ries–Smith data in Table

5-1, using the values of $G^2$ listed in Table 5-3 and other values given by Goodman [1971a]. We again use a 0.05 level of significance for illustrative purposes. First we note that

$$[1][2][3][4] \tag{5.8}$$

does not fit the data, but that

$$[12][13][14][23][24][34] \tag{5.9}$$

and

$$[123][124][134][234] \tag{5.10}$$

do. Thus we work to find a model that lies between (5.8) and (5.9).

Using forward selection we begin with (5.8) and add $u_{24}$, since the difference between the $G^2$ values for (5.8) and for

$$[1][3][24] \tag{5.11}$$

is $42.9 - 22.4 = 20.5$ with 1 d.f. (significant at the 0.05 level), and this is the most significant $u$-term we can add. Next we add $u_{34}$, the most significant of the remaining two-factor $u$-terms, based on the difference between the $G^2$ values for (5.11) and for

$$[1][24][34], \tag{5.12}$$

which is $22.4 - 18.0 = 4.4$ with 1 d.f. At this point we are unable to delete any terms. Now we add $u_{13}$, the next most significant term, based on the difference between the $G^2$ values for (5.12) and for

$$[13][24][34], \tag{5.13}$$

which is $18.0 - 11.9 = 6.1$ with 2 d.f. We can neither delete nor add any further two-factor terms, and so the forward selection procedure leads us to (5.13) as the "best" model.

Using backward elimination we begin with (5.9) and delete $u_{14}$, since the difference between $G^2$ values for (5.9) and for

$$[12][13][23][24][34] \tag{5.14}$$

is $10.1 - 9.9 = 0.2$ with 2 d.f. Next we drop $u_{12}$, based on the difference between $G^2$ values for (5.14) and for

$$[13][23][24][34], \tag{5.15}$$

which is $11.2 - 10.1 = 1.1$ with 2 d.f. We are unable to add back in any terms at this stage, and so we proceed to delete $u_{23}$, based on the difference between

$G^2$ values for (5.15) and for (5.13), which is $11.9 - 11.2 = 0.7$ with 1 d.f. Since we are unable to add or delete any more terms, we again end up with (5.13) as the "best" model.

Just as with the corresponding methods used in regression analysis, it is possible for forward selection and backward elimination to lead to different "best" models. As with stepwise regression procedures (see Draper and Smith [1966]), the significance values of the tests carried out should be interpreted with great caution because the test statistics are often highly dependent.

Goodman [1971a] suggests abbreviated stepwise procedures employing methods similar to the ones described above, along with asymptotic variance estimates for the $u$-terms based on the general loglinear model with no terms set equal to zero. Wermuth [1976a, b], using analogies between certain loglinear models for contingency tables and certain covariance selection models, proposes a noniterative backward selection procedure based initially only on models involving direct estimates. Benedetti and Brown [1976] compare several strategies of model selection, including the stepwise methods of Goodman but not that of Wermuth, in the context of two examples, and they compare the results in terms of whether an appropriate model is found and the total number of models that must be fit before it is found.

It must be understood that the various stepwise methods require a considerable amount of computation, and they should not be thought of as automatic devices for deciding upon appropriate loglinear models. At best they can be of aid in limiting attention to a reduced set of models that give a reasonable fit to the data. In fact other, perhaps more parsimonious and substantively interesting, models might exist that also provide an acceptable fit.

## 5.4 Looking at All Possible Effects

One of the advantages of the conventional analysis of variance for quantitative data, in which all estimable contrasts are entered, is that it forces us to look at effects that we might be tempted to ignore or assume away. Within the framework of loglinear models it is still possible to look at all possible effects or parameters, and at least two approaches seem reasonable.

Brown [1976] proposes that the importance of effects in loglinear models be studied by computing two ad hoc test statistics for each effect. Suppose we have a four-dimensional table and are looking at the effect $u_{123}$. Brown would first test for the absence of $u_{123}$ in the loglinear model for the marginal table involving variables 1, 2, and 3,

(1) [12] [13] [23].
Then he would test the fit of the model for the full table that includes all interactions of order 3,
(2) [123] [124] [134] [234],
and the fit of the model with all interactions of order 3 except $u_{123}$,
(3) [124] [134] [234].
The marginal-likelihood-ratio test for $u_{123}$, $G^2(1)$, and the conditional test for the absence of $u_{123}$, $G^2(3) - G^2(2)$, both have the same degrees of freedom and are often indicative of the magnitude of other conditional tests for the same term. Thus when both statistics are small, it is unlikely that other conditional test statistics for that effect will be large, and the effect would appear not to be needed in the model. When both are large, it is unlikely that other conditional test statistics will be small, so the interaction term would appear to be required in the model. When the two statistics are discrepant, we may need to consider further models with and without the effect in question.

In a large number of examples this type of screening approach leads to the consideration of a limited number of loglinear models as part of a second step in the model-building process. Since the two test statistics used for a given effect do not bound all possible values of the conditional test statistics for that effect, Brown's screening approach is fallible, and it should be used with caution.

Table 5-5 contains the values $G^2(1)$ and $G^2(2) - G^2(3)$ for all interaction $u$-terms computed by Brown [1976] using the Ries–Smith data from Table 5-1. A cursory examination of Table 5-5 suggests that $u_{24}$ must definitely be

**Table 5-5**
Marginal and Conditional Tests for Interaction Effects for the Data in Table 5-1

| $u$-term | d.f. | Marginal $G^2$ | $G^2(2) - G^2(3)$ |
|---|---|---|---|
| $u_{12}$ | 2 | 1.1 | 1.0 |
| $u_{13}$ | 2 | 6.1 | 6.1 |
| $u_{14}$ | 2 | 0.4 | 0.2 |
| $u_{23}$ | 1 | 1.3 | 0.7 |
| $u_{24}$ | 1 | 20.6 | 19.9 |
| $u_{34}$ | 1 | 4.4 | 3.7 |
| $u_{123}$ | 2 | 1.6 | 1.4 |
| $u_{124}$ | 2 | 5.3 | 4.6 |
| $u_{134}$ | 2 | 0.1 | 0.2 |
| $u_{234}$ | 1 | 2.8 | 2.2 |
| $u_{1234}$ | 2 | 0.7 | 0.7 |

included in our model, and that $u_{34}$ and $u_{13}$ require further consideration. This screening procedure leads directly to a secondary analysis equivalent to that presented earlier in Section 5.2.

A second approach to looking at all possible effects relies on reasonably simple formulas for the asymptotic or large-sample variances of estimated $u$-terms in the full or saturated model. What makes this approach work is the fact that each subscripted $u$-term in the general loglinear model can be expressed as a linear combination of the logarithms of the expected values (or equivalently the logarithms of the cell probabilities), where the weights or coefficients used in the linear combination add to zero. Such linear combinations are referred to as *linear contrasts.* For example, if we set $Z_{ijkl} = \log p_{ijkl}$ or $Z_{ijkl} = \log m_{ijkl}$ in a four-dimensional table, then

$$\begin{aligned} u_{1234(ijkl)} = {} & Z_{ijkl} - \frac{1}{I}Z_{+jkl} - \frac{1}{J}Z_{i+kl} - \frac{1}{K}Z_{ij+l} - \frac{1}{L}Z_{ijk+} \\ & + \frac{1}{IJ}Z_{++kl} + \frac{1}{IK}Z_{+j+l} + \frac{1}{IL}Z_{+jk+} + \frac{1}{JK}Z_{i++l} + \frac{1}{JL}Z_{i+k+} \\ & + \frac{1}{KL}Z_{ij++} - \frac{1}{IJK}Z_{+++l} - \frac{1}{IJL}Z_{++k+} - \frac{1}{IKL}Z_{+j++} \\ & - \frac{1}{JKL}Z_{i+++} + \frac{1}{IJKL}Z_{++++}, \end{aligned} \tag{5.16}$$

and the right-hand side of (5.16) can then be rewritten as a linear contrast of the $\{Z_{ijkl}\}$, that is, as

$$u_{1234(ijkl)} = \sum_{i,j,k,l} \beta_{ijkl} Z_{ijkl}, \tag{5.17}$$

where

$$\sum_{i,j,k,l} \beta_{ijkl} = 0. \tag{5.18}$$

When $I = J = K = L = 2$, there is only one parameter corresponding to each $u$-term, and expression (5.16) reduces to

$$\begin{aligned} 16u_{1234(1111)} = {} & Z_{1111} + Z_{2211} + Z_{1221} + Z_{1122} + Z_{2121} + Z_{2112} \\ & + Z_{1212} + Z_{2222} - Z_{2111} - Z_{1211} - Z_{1121} - Z_{1112} \\ & - Z_{1222} - Z_{2122} - Z_{2212} - Z_{2221}. \end{aligned} \tag{5.19}$$

Other $u$-terms for the $2 \times 2 \times 2 \times 2$ table have the same form with coefficients $\beta_{ijkl} = \pm \frac{1}{16}$, just as in the conventional analysis of variance.

The MLE of a linear contrast of the form (5.17), subject to (5.18), is found by substituting $\log x_{ijkl}$ for $Z_{ijkl}$:

$$\hat{u}_{1234(ijkl)} = \sum_{i,j,k,l} \beta_{ijkl} \log x_{ijkl}. \tag{5.20}$$

The large-sample variance of the estimated contrast (5.20) is

$$\sum_{i,j,k,l} \beta_{ijkl}^2 m_{ijkl}^{-1},$$

and this variance can be consistently estimated by

$$\sum_{i,j,k,l} \beta_{ijkl}^2 x_{ijkl}^{-1}. \tag{5.21}$$

For large samples, a linear contrast of the form (5.17) approximately follows a normal distribution with mean

$$\sum_{i,j,k,l} \beta_{ijkl} \log p_{ijkl} = \sum_{i,j,k,l} \beta_{ijkl} \log m_{ijkl} \tag{5.22}$$

if the $\{x_{ijkl}\}$ follow a Poisson, multinomial, or product-multinomial sampling model. In the latter case, some linear contrasts are fixed by design, so that it makes no sense to estimate them. In the $2 \times 2 \times 2 \times 2$ table, the estimated $u$-terms in the saturated model all have the same estimated large-sample variance,

$$\left(\frac{1}{16}\right)^2 \sum_{i,j,k,l} x_{ijkl}^{-1}. \tag{5.23}$$

Since we would like to look at several $u$-terms simultaneously, we need to know something about the joint distribution of two or more contrasts. The following theorem is a direct application of the $\delta$ method (see Bishop, Fienberg, and Holland [1975]), and the result has been used by various authors (e.g., Fienberg [1969], Goodman [1964, 1971a], Lindley [1964]). The result is applicable for any number of dimensions, and thus we give it using single subscripts.

**Theorem 5-1. Suppose that the $t$ counts $\{x_i\}$ have a multinomial distribution with corresponding cell probabilities $\{p_i\}$ and total sample size $N$, and that the constants $\{\beta_i^{(q)}\}$ satisfy $\Sigma_i \beta_i^{(q)} = 0$, where $\beta_i^{(q)} \neq 0$ for some $i$ and $q = 1, 2, \ldots, t^*$ with $t^* < t$. Then the joint distribution of the estimated loglinear contrasts,**

$$\sum_i \beta_i^{(q)} \log x_i, \qquad q = 1, 2, \ldots, t^*,$$

**is approximately multivariate normal with means**

$$\sum_i \beta_i^{(q)} \log p_i = \sum_i \beta_i^{(q)} \log m_i, \qquad q = 1, 2, \ldots, t^*,$$

**and covariances (variances where $s = q$)**

$$\sum_i \beta_i^{(q)} \beta_i^{(s)} (Np_i)^{-1} = \sum_i \beta_i^{(q)} \beta_i^{(s)} m_i^{-1}, \qquad q, s = 1, 2, \ldots, t^*,$$

**which can be consistently estimated by**

$$\sum_i \beta_i^{(q)} \beta_i^{(s)} x_i^{-1}, \qquad q, s = 1, 2, \ldots, t^*.$$

We can now make use of Theorem 5-1 by getting estimates of each $u$-term (substituting $\log x_i$ for $Z_i = \log m_i$ in the appropriate loglinear contrast) and then dividing the estimated $u$-term by its estimated large-sample standard deviation. Goodman [1964] has discussed the simultaneous-inference problem for these standardized estimated $u$-terms (note that the covariances among the estimated $u$-terms are typically nonzero).

Bishop [1969], Goodman [1970], and others have reanalyzed data presented by Dyke and Patterson [1952] based on an observational study by Lombard and Doering [1947], in which a sample of 1729 individuals are cross-classified in a $2^5$ table according to whether they (1) read newspapers, (2) listen to the radio, (3) do "solid" reading, (4) attend lectures, and (5) have good or poor knowledge regarding cancer. Dyke and Patterson chose to view variable 5 (knowledge) as a response variable and the other four variables as explanatory, but following the suggestion of Bishop [1969], we do not make such distinctions here at this time. Table 5-6 presents the actual data, and Table 5-7 lists the estimated $u$-terms and their standardized values (i.e., estimated $u$-term divided by estimated standard deviation). We note that the values in Table 5-7 are not identical to those given by Goodman [1970], who added 0.5 to each cell count before computing the estimated $u$-terms and their standard deviations. Since we are dealing with a $2^5$ table, the estimated standard deviation for each $u$-term is the same:

$$\frac{1}{32}\left( \sum_{i,j,k,l,m} x_{ijklm}^{-1} \right)^{1/2} = 0.0647.$$

**Table 5-6**
Data from Dyke and Patterson [1952]

| | | Radio | | | | No Radio | | | |
|---|---|---|---|---|---|---|---|---|---|
| | | Solid Reading | | No Solid Reading | | Solid Reading | | No Solid Reading | |
| | | Knowledge | | | | Knowledge | | | |
| | | Good | Poor | Good | Poor | Good | Poor | Good | Poor |
| Newspaper | Lectures | 23 | 8 | 8 | 4 | 27 | 18 | 7 | 6 |
| | No Lectures | 102 | 67 | 35 | 59 | 201 | 177 | 75 | 156 |
| No Newspaper | Lectures | 1 | 3 | 4 | 3 | 3 | 8 | 2 | 10 |
| | No Lectures | 16 | 16 | 13 | 50 | 67 | 83 | 84 | 393 |

Perhaps the simplest way to examine the standardized estimated $u$-terms is graphically. Cox and Lauh [1967], for a similar problem, suggest plotting the standardized values on half-normal probability paper (see Daniel [1959]), even though such a graph ignores the correlations among the standardized estimated $u$-terms. Figure 5-1 shows such a plot for the standardized terms in Table 5-7. The standardized values corresponding to $u_4$, $u_2$, $u_1$, $u_{13}$, $u_{15}$, and $u_{24}$ all exceed 3.00 in absolute value; we exclude these from the plot and work with the remaining 25 subscripted $u$-terms.

The 25 points in Figure 5-1 all lie roughly along a straight line, with the exception of the ones corresponding to $u_{12}$ and $u_5$. Combining these two terms with the six excluded from the plot, we are led to consider the model

$$u + u_1 + u_2 + u_3 + u_4 + u_5 + u_{12} + u_{13} + u_{15} + u_{24}. \tag{5.24}$$

Fitting this model to the data in Table 5-5 yields goodness-of-fit values of $X^2 = 151.2$ and $G^2 = 144.1$ with 22 d.f. The fit of the model is quite poor, and an improved fit might be achieved by including the $u$-term corresponding to the next largest standardized value, $u_{345}$. Since we are restricting our attention to hierarchical models, our new model is

$$u + u_1 + u_2 + u_3 + u_4 + u_5 + u_{12} + u_{13} + u_{15} + u_{24} + u_{34} + u_{35} + u_{45} + u_{345}. \tag{5.25}$$

For this model, $X^2 = 26.9$ and $G^2 = 27.2$ with 18 d.f., values corresponding to a descriptive level of significance slightly greater than 0.05. Adding the term corresponding to the next highest absolute standardized value, $u_{25}$, yields a markedly improved fit: $X^2 = 18.5$ and $G^2 = 20.1$ with 17 d.f. Thus the simplest model that fits the data reasonably well is

**Table 5-7**
Estimated $u$-terms and Standardized Values for Data of Dyke and Patterson [1952] Given in Table 5-6

| $u$-term | Estimated Value Based on Saturated Model | Standardized Value | Absolute Rank | Estimated Value Based on Model (5.24) |
|---|---|---|---|---|
| $u_1$ | 0.441 | 6.82 | 3 | −0.343 |
| $u_2$ | −0.444 | −6.86 | 2 | −0.486 |
| $u_3$ | 0.081 | 1.25 | 17 | 0.094 |
| $u_4$ | −1.219 | −18.84 | 1 | −1.161 |
| $u_5$ | −0.153 | −2.36 | 8 | −0.098 |
| $u_{12}$ | 0.155 | 2.39 | 7 | 0.235 |
| $u_{13}$ | 0.326 | 5.04 | 4 | 0.363 |
| $u_{14}$ | 0.112 | 1.73 | 12 | — |
| $u_{15}$ | 0.253 | 3.91 | 5 | 0.164 |
| $u_{23}$ | −0.042 | −0.65 | 26 | — |
| $u_{24}$ | 0.205 | 3.17 | 6 | 0.176 |
| $u_{25}$ | 0.121 | 1.87 | 11 | 0.080 |
| $u_{34}$ | 0.106 | 1.64 | 13 | 0.171 |
| $u_{35}$ | 0.135 | 2.09 | 10 | 0.159 |
| $u_{45}$ | 0.085 | 1.31 | 16 | 0.149 |
| $u_{123}$ | −0.003 | −0.05 | 31 | — |
| $u_{124}$ | −0.071 | −1.10 | 18 | — |
| $u_{125}$ | −0.016 | −0.25 | 28.5 | — |
| $u_{134}$ | 0.012 | 0.19 | 30 | — |
| $u_{135}$ | 0.016 | 0.25 | 28.5 | — |
| $u_{145}$ | 0.104 | 1.61 | 14 | — |
| $u_{234}$ | −0.100 | −1.55 | 15 | — |
| $u_{235}$ | −0.056 | −0.87 | 24 | — |
| $u_{245}$ | 0.064 | 0.99 | 21.5 | — |
| $u_{345}$ | −0.144 | −2.23 | 9 | −0.101 |
| $u_{1234}$ | 0.051 | 0.79 | 25 | — |
| $u_{1235}$ | 0.068 | 1.05 | 20 | — |
| $u_{1245}$ | −0.021 | −0.32 | 27 | — |
| $u_{1345}$ | 0.070 | 1.08 | 19 | — |
| $u_{2345}$ | −0.063 | −0.97 | 23 | — |
| $u_{12345}$ | 0.064 | 0.99 | 21.5 | — |

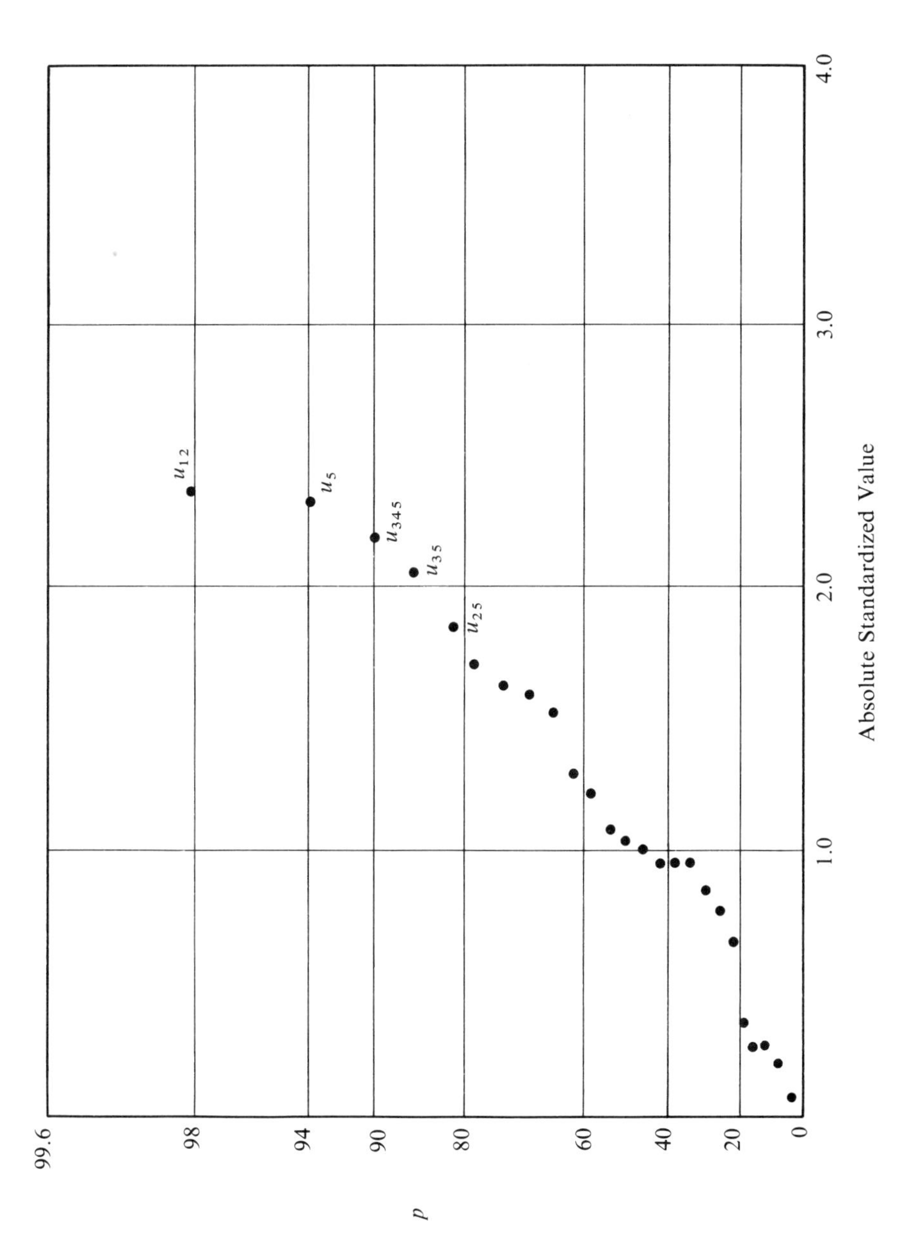


**Figure 5-1**
Half-Normal Plot of Absolute Standardized Values from Table 5-7 (Excluding Six Largest Values)

$$
\begin{aligned}
& u + u_1 + u_2 + u_3 + u_4 + u_5 + u_{12} + u_{13} \\
& + u_{15} + u_{24} + u_{25} + u_{34} + u_{35} + u_{45} + u_{345}.
\end{aligned}
\tag{5.26}
$$

In Chapter 6 an interpretation of (5.26) is discussed in which knowledge (variable 5) is viewed as a response variable and the remaining four variables are viewed as explanatory. The final column of Table 5-7 lists the MLEs of the $u$-terms in model (5.26). These estimates are in some instances markedly different from those based on the saturated model, which are given in the second column of the table. Moreover, the large-sample standard deviation used to produce the standardized values in the table is no longer appropriate for the new estimated $u$-terms. This value, 0.0647, is now an overestimate of the large-sample standard deviation associated with the new estimates. For more details on the computation of asymptotic standard deviations, see Haberman [1974a], Bock [1975], Koch, Freeman, and Tolley [1975], Ku and Kullback [1974], and Lee [1975], and also Appendix II of this book. Most computer programs designed to fit loglinear models to multidimensional tables allow the user to request the estimated $u$-terms and their estimated standard errors.

A sensible approach to looking at all possible effects would probably use a combination of the two methods described in this section, for a restricted set of $u$-terms.

# 6
# Fixed Margins and Logit Models

In Chapters 2, 3, and 5 we noted that the estimated expected cell values for loglinear models under product-multinomial sampling (with a set of fixed marginal totals) are the same as those under multinomial sampling, provided the $u$-terms corresponding to the fixed margins are included in the model. We give some details regarding this point here, and then we turn to a class of models closely related to loglinear models.

## 6.1 A Three-Dimensional Example

Wakeley [1954] investigated the effect of planting longleaf and slash pine seedlings $\frac{1}{2}$ inch too high or too deep in winter upon their mortality the following fall. The data, reported by Bliss [1967], are reproduced in Table 6-1a. The row total for each type of seedling is fixed at 100 by the experimental design.

We can still fit loglinear models to the data in 6-1a, but we must include the $u_{13}$-term in all of the models to be considered. In the iterative proportional fitting procedure, this will keep the estimated expected two-dimensional margin totals for variables 1 and 3 (depth of planting and seedling type) equal to the observed totals fixed by design. In interpreting these fitted models, however, we usually do not speak of the two-factor effect relating depth and type.

**Table 6-1**
Effect of Depth of Planting on the Mortality of Pine Seedlings (Wakeley [1954])

(a) Observed data

| | Longleaf Seedlings | | | Slash Seedlings | | |
|---|---|---|---|---|---|---|
| Depth of Planting | Dead | Alive | Totals | Dead | Alive | Totals |
| Too high | 41 | 59 | 100 | 12 | 88 | 100 |
| Too low | 11 | 89 | 100 | 5 | 95 | 100 |
| Totals | 52 | 148 | 200 | 17 | 183 | 200 |

(b) Expected values ($u_{123} = 0$)

| Depth of Planting | Dead | Alive | Totals | Dead | Alive | Totals |
|---|---|---|---|---|---|---|
| Too high | 39.38 | 60.62 | 100 | 13.62 | 86.38 | 100 |
| Too low | 12.62 | 87.38 | 100 | 3.38 | 96.62 | 100 |
| Totals | 52 | 148 | 200 | 17 | 183 | 200 |

**Table 6-2**
Loglinear Models Fit to Data in Table 6-1, and their Goodness-of-Fit Statistics (All Models Include $u_{13}$)

| Model | $X^2$ | $G^2$ | d.f. |
|---|---|---|---|
| [12] [13] [23] | 1.37 | 1.28 | 1 |
| [13] [23] | 26.54 | 27.79 | 2 |
| [12] [13] | 24.03 | 25.03 | 2 |
| [13] [2] | 54.70 | 50.10 | 3 |

Table 6-2 displays the permissible models for this set of data and gives the values of the corresponding goodness-of-fit statistics. Clearly only the no three-factor effect model fits the data at all well, and the estimated expected cell values under this model are given in Table 6-1b. (Had we fitted the model that also sets $u_{13} = 0$, we would have found small values of $X^2$ and $G^2$, but the estimated expected marginal totals for variables 1 and 3 would no longer be equal to 100.)

How do we interpret this fitted model? Since we are interested in the effects of depth and seedling type on mortality, it is reasonable for us to look at the mortality ratio (no. dead/no. alive) for each combination of depth and seedling type, that is, $m_{i1k}/m_{i2k}$ for all $i$ and $k$. The fitted model says that

$$\log m_{ijk} = u + u_{1(i)} + u_{2(j)} + u_{3(k)} + u_{12(ij)} + u_{13(ik)} + u_{23(jk)} \tag{6.1}$$

and thus

$$\begin{aligned}\log\left(\frac{m_{i1k}}{m_{i2k}}\right) &= [u_{2(1)} - u_{2(2)}] + [u_{12(i1)} - u_{12(i2)}] + [u_{23(1k)} - u_{23(2k)}] \\ &= 2[u_{2(1)} + u_{12(i1)} + u_{23(1k)}] \\ &= w + w_{1(i)} + w_{3(k)},\end{aligned} \tag{6.2}$$

where in the last line of (6.2) the subscript indicating variable 2 (mortality) is suppressed since the left-hand side is just the log mortality ratio. Model (6.2) is usually referred to as a linear *logit* model (see, for example, Bishop [1969]); it says that there are additive effects on the log mortality ratio due to depth and to seedling type, but that there is no two-factor effect of depth and seedling type considered jointly. Note that the $u_{13}$-term, which we had to include in the loglinear model, does not appear in the resulting logit model for the mortality ratios. This is one reason why we need not have been concerned by its forced inclusion. Had we included the three-factor effect $u_{123}$ in the loglinear model (6.1), we would have had a two-factor term $w_{13(ik)} (= 2u_{123(i1k)})$ in the logit model (6.2).

## 6.2 Logit Models

When one is interested in models that assess the effects of categorical variables on a dichotomous "response" variable, one is often led to the formation of logit models for the response variable, especially if all other variables are "design" variables that one wishes to manipulate. As we have just seen, logit models contain terms corresponding to those in loglinear models. Many investigators fit logit models to data sets in which the remaining nonresponse variables are not design variables, and thus they lose information on the relationships among the nonresponse variables (see Bishop [1969]). If one's interest lies only in the effects of the other (explanatory) variables on the response variable, however, then an analysis based on logits for the response variable is appropriate.

We illustrate this point using the Ries–Smith data from Table 5-1. We view the variable "preference" as the response variable and we pretend to be interested only in how "softness," "use," and "temperature" affect "preference." Thus the general logit model for this problem is

$$\text{logit}_{ijk} = \log\left(\frac{m_{ijk1}}{m_{ijk2}}\right) \tag{6.3}$$

$$= w + w_{1(i)} + w_{2(j)} + w_{3(k)} + w_{12(ij)} + w_{23(jk)} + w_{13(ik)} + w_{123(ijk)},$$

with the usual ANOVA-like constraints on the $w$-terms. In order to fit a logit model that is a special case of expression (6.3) using iterative proportional fitting, we must fit a loglinear model in which the terms $u_1, u_2, u_3, u_{12}, u_{13}, u_{23}$, and $u_{123}$ are all included; that is, the model must include [123]. This is so because we are taking as fixed the three-dimensional marginal totals corresponding to the explanatory variables. For example, if we fit the loglinear model

$$[123]\,[234], \tag{6.4}$$

then we can translate our results directly into those for the logit model

$$w + w_{2(j)} + w_{3(k)} + w_{23(jk)}, \tag{6.5}$$

and estimated expected values for (6.4) can be used to get the estimated logits for model (6.5).

For the Ries–Smith data, Goodman [1971a] shows that the best-fitting (i.e., giving the best fit with the fewest number of parameters) logit model based on logits for the preference variable is of the form

$$w + w_{2(j)}, \tag{6.6}$$

which corresponds to the loglinear model

$$[123][24].$$

The three-dimensional totals corresponding to softness–use–temperature were not fixed by design. The reason we consider a logit model for the data is that we view preference as a response variable and softness, use, and temperature as explanatory variables. As a result, we condition on the three-dimensional totals involving the explanatory variables. When this distinction between response and explanatory variables is not important the loglinear model approach does not require conditioning on these totals and may well yield better estimates for parameters of particular interest. (For a detailed discussion on the advantages of loglinear models over logit models in such circumstances, see Bishop [1969].)

For the Dyke–Patterson data of Table 5-6, our final loglinear model was

$$[12][13][15][24][25][345]. \tag{6.7}$$

If we treat variable 5 (knowledge of cancer) as a response variable and condition on the remaining four explanatory variables, we get the logit model

$$w + w_{1(i)} + w_{2(j)} + w_{3(k)} + w_{4(l)} + w_{34(kl)}. \tag{6.8}$$

This fitted model implies that exposure to newspapers, radio, solid reading, and lectures all have a positive effect on knowledge of cancer, but that there is also a joint effect of solid reading and exposure to lectures that is negative. The combined effect of these two explanatory variables is not as great as the sum of their individual effects, but it is greater than the effect of either one separately.

Cox [1970a] and others suggest that one should always condition on the explanatory variables, whether they are fixed by design or not. For the Dyke–Patterson data such an approach would lead to fitting the model

$$[15][25][345][1234] \tag{6.9}$$

rather than the one given by expression (6.7). This loglinear model produces the same logit model as before (i.e., (6.8)), but the estimates of the parameters differ somewhat between the two fitted logit models, although the differences in this example are small (see Bishop [1969] for a detailed comparison). The difficulty with this conditional approach comes when the data are spread thinly over the multiway cross-classification and the marginal total fixed by conditioning on the explanatory variables contains observed zero entries. The estimated expected values for cells corresponding to these zero marginal entries must then be zero, and the corresponding logits are undefined. For example, suppose the (1, 1, 2, 2, 1) and (1, 1, 2, 2, 2) cells in the Dyke–Patterson data contain zero entries. Then the (1, 1, 2, 2, +) margin total is zero, and we get zero estimated expected values for the cells that add up to this marginal total if we fit model (6.9).

There are two ways out of such a dilemma. The first approach is to fit a loglinear model to the full array but condition only on marginal totals that are fixed by design. The problematic zero marginal entries may then disappear as a result of smoothing. In our hypothetical modification to the Dyke–Patterson data, with zeros replacing the entries in the (1, 1, 2, 2, 1) and (1, 1, 2, 2, 2) cells, if we fit the loglinear model corresponding to the logit model of expression (6.8), the estimated expected values for these cells would be nonzero. The work of Bishop and Mosteller [1969] on the National Halothane Study illustrates the strengths of this approach.

The second way to handle the dilemma posed by problematic marginal zero entries in logit models is to simplify the hypothesized relationship between the response and explanatory variables. When the explanatory variables are polytomous, we might take advantage of the implicit ordering of categories and consider fitting logit models in which the effects of explanatory variables are represented by linear regression-like coefficients. Such an approach is described in more detail in the next section.

Logit models are the categorical response analogs to regression models for continuous response variates. Nelder and Wedderburn [1972] and Nelder [1974] describe a general technique of iterative weighted linear regression that is applicable to both the regular regression model with normal errors and the logit model, as well as many others. The methods described here are more limited. DuMouchel [1975], by analogy with the $R^2$ criterion in standard regression problems, has proposed a particular measure of the predictive error of a logit model, but it seems somewhat premature to latch onto a specific measure for this purpose, and thus DuMouchel's measure will not be described here. It is important, however, to consider the predictive ability of logit models in many situations.

## 6.3 Logit Models and Ordered Categories

In Section 4.4 our results on loglinear models were extended to allow for structured interaction $u$-terms that directly reflect information regarding scores associated with ordered categories. Such models provide a useful tool for making the transition from the logit models described above to the linear logistic response models to be discussed in the next section.

Suppose we have a $2 \times J \times K$ table, with variable 1 being a binary response variable and variables 2 and 3 being explanatory variables with ordered categories, the scores for which are $\{v_j^{(2)}\}$ and $\{v_k^{(3)}\}$, respectively. Furthermore, suppose that in the loglinear model being considered, two-factor effects relating the explanatory variables to the response variable reflect the ordered structure as in expression (4.24), so that

$$u_{12(ij)} = (v_j^{(2)} - \bar{v}^{(2)})u_{1(i)}^{(2)} \tag{6.10}$$

and

$$u_{13(ik)} = (v_k^{(3)} - \bar{v}^{(3)})u_{1(i)}^{(3)}. \tag{6.11}$$

The hierarchical restriction on ordered effects then implies that

$$u_{123(ijk)} = (v_j^{(2)} - \bar{v}^{(2)})(v_j^{(3)} - \bar{v}^{(3)})u_{1(i)}^{(23)}. \tag{6.12}$$

The terms $u_{1(i)}$, $u_{1(i)}^{(2)}$, $u_{1(i)}^{(3)}$, and $u_{1(i)}^{(23)}$ may all be different. Finally, we must include the $u_{23}$-term in the loglinear model since variable 2 and 3 are explanatory, and thus we need to condition on the [23] marginal total.

Now we wish to translate the loglinear model just described into a logit model for the response variable. For simplicity we set $u_{1(i)}^{(23)} = 0$. If we set

$$\beta_2 = [u_{1(1)}^{(2)} - u_{1(2)}^{(2)}] = 2u_{1(1)}^{(2)}, \tag{6.13}$$

$$\beta_3 = 2u_{1(1)}^{(3)}, \tag{6.14}$$

and

$$\beta_0 = 2u_{1(1)} - \beta_2\bar{v}^{(2)} - \beta_3\bar{v}^{(3)}, \tag{6.15}$$

then we have

$$\text{logit}_{jk} = \log \frac{m_{1jk}}{m_{2jk}} = \beta_0 + \beta_2 v_j^{(2)} + \beta_3 v_k^{(3)}. \tag{6.16}$$

This is a linear regression-like model for the logits based on the binary response variable.

Using the results of Section 4.4 for our standard sampling models, we have the following likelihood equations:

$$\hat{m}_{+jk} = x_{+jk}, \qquad j = 1, 2, \ldots, J, \quad k = 1, 2, \ldots, K, \tag{6.17}$$

$$\sum_j v_j^{(2)}\hat{m}_{ij+} = \sum_j v_j^{(2)}x_{ij+}, \qquad i = 1, 2, \tag{6.18}$$

$$\sum_k v_k^{(3)}\hat{m}_{i+k} = \sum_k v_k^{(3)}x_{i+k}, \qquad i = 1, 2, \tag{6.19}$$

and

$$\hat{m}_{i++} = x_{i++}, \qquad i = 1, 2. \tag{6.20}$$

Equations (6.18) and (6.19) resemble the normal equations of standard regression theory, equation (6.17) corresponds to binary response structure, and equation (6.20) guarantees that the expected number of observations equals the observed, at each level of the response variable. As in Section 4.4, we could solve these equations for the $\{\hat{m}_{ijk}\}$ using a version of iterative proportional fitting, or some other iterative method such as weighted nonlinear least-squares (see Jennrich and Moore [1975] or Nelder and Wedderburn [1972]) or the Newton–Raphson algorithm (see, for example, Haberman [1974a]).

Cox [1970a] describes a problem involving the preparation and testing of ingots. For each of four heating times and five soaking times a set of ingots was to be tested. Heating time and soaking time are explanatory variables, and the response variable indicates whether or not the ingot is ready for rolling. Thus we have a $2 \times 4 \times 5$ array in which the values associated with the categories of the explanatory variables are simply the heating times and soaking times:

$$v^{(2)} = \begin{pmatrix} 7 \\ 14 \\ 27 \\ 51 \end{pmatrix} \quad \text{and} \quad v^{(3)} = \begin{pmatrix} 1.0 \\ 1.7 \\ 2.2 \\ 2.8 \\ 4.0 \end{pmatrix}.$$

The observed values are given in Table 6-3. Note that there were tests for only 19 of the 20 combinations of heating and soaking time.

We begin our analysis by testing the fit of the no second-order interaction models $u_{123} = 0$: $G^2 = 11.3$ and $X^2 = 9.7$ with 7 d.f., values corresponding to a descriptive level of significance greater than 0.20. Note that the d.f. are not equal to $(I - 1)(J - 1)(K - 1) = 12$. This is a result of two zero marginal totals, $x_{11+} = 0$ and $x_{+44} = 0$, which imply that seven interior cells will have zero expected cell values. We adjust the d.f. to take this fact into account; thus d.f. $= 12 - 7 + 2 = 7$. For further discussion of how to deal with marginal zeros, see Chapter 8.

Next we fit the logit model based on $u_{123} = 0$ and expressions (6.10) and (6.11): $G^2 = 13.8$ with 11 d.f., corresponding to a descriptive level of significance considerably greater than 0.05. This model fits the data reasonably well and provides a significant improvement over the no second-order interaction model at the 0.05 level of significance. Since $\hat{\beta}_3$ is small relative to its estimated large-sample standard deviation, we consider the model with

**Table 6-3**
Data on Ingots for Various Combinations of Heating Time and Soaking Time* (Cox [1970])

| | | Heating Time | | | | |
|---|---|---|---|---|---|---|
| | | 7 | 14 | 27 | 51 | Totals |
| | 1.0 | 0, 10 | 0, 31 | 1, 55 | 3, 10 | 4, 106 |
| Soaking | 1.7 | 0, 17 | 0, 43 | 4, 40 | 0, 1 | 4, 101 |
| Time | 2.2 | 0, 7 | 2, 31 | 0, 21 | 0, 1 | 2, 60 |
| | 2.8 | 0, 12 | 0, 31 | 1, 21 | 0, 0 | 1, 64 |
| | 4.0 | 0, 9 | 0, 19 | 1, 15 | 0, 1 | 1, 44 |
| Totals | | 0, 55 | 2, 155 | 7, 152 | 3, 13 | 12, 375 |

* First figure in each cell is number not ready for rolling, and second figure is number ready for rolling.

$\beta_3 = 0$. It also provides an acceptable fit to the data, and it is the one Cox [1970a] deems appropriate for the data. For this model $\hat{\beta}_2 = 0.0807$. Notice how the model can be used to get a predicted logit value for the zero marginal total corresponding to a soaking time of 2.8 and a heating time of 51.

## 6.4 Linear Logistic Response Models

Continuing with the problem of the preceding section, we let $J$ and $K$ increase in size to produce an ever finer grid of values for the two explanatory variables. As $J$ and $K$ increase we approach a situation in which the explanatory variables are treated as continuous rather than categorical. If we keep the sample size $N = x_{+++}$ fixed while increasing $J$ and $K$, then in the most extreme situation we would expect to end up with values of $x_{+jk}$ that are either zero or one. If we continue to view this as a contingency table problem, the standard large-sample theory no longer applies since the size of "the table" increases with the sample size $N$. Similar large-sample results for the parameters of interest are available, however (see, for example, Haberman [1974a]).

An alternate, but equivalent, way to describe this situation is as follows. Let $y_1, y_2, \ldots, y_N$ be independent binary random variables taking values 0 or 1, with

$$\Pr(y_i = 1) = p_i, \qquad i = 1, 2, \ldots, N. \tag{6.21}$$

For the $i$th such variable we have observations $z_{i1}$ and $z_{i2}$ on the two explanatory variables, and

$$\log\left(\frac{p_i}{1 - p_i}\right) = \beta_0 + \beta_1 z_{i1} + \beta_2 z_{i2}, \qquad i = 1, 2, \ldots, N. \tag{6.22}$$

The general linear logistic model, with $r$ explanatory variables and a binary response variable, resembles the two-variable model above. Let $y_1$, $y_2, \ldots, y_N$ be independent binary random variables taking values 0 or 1, with

$$\Pr(y_i = 1) = p_i = \frac{\exp(\beta_0 + \Sigma_{j=1}^r \beta_j z_{ij})}{1 + \exp(\beta_0 + \Sigma_{j=1}^r \beta_j z_{ij})}, \qquad i = 1, 2, \ldots, N. \tag{6.23}$$

The quantities $z_{ij}$ for $j = 1, 2, \ldots, r$ are the values of the $r$ explanatory variables for the $i$th observation and are considered fixed. Expression (6.23) is equivalent to a linear model in the logits:

$$\log\left(\frac{p_i}{1 - p_i}\right) = \beta_0 + \sum_{j=1}^{r} \beta_j z_{ij}, \qquad i = 1, 2, \ldots, N. \tag{6.24}$$

In this formulation the explanatory variables may be binary, categorical but with explicit orderings, as in Section 6.3, or continuous.

The minimal sufficient statistics for the linear logistic regression model (6.23) are

$$\sum_{i=1}^{N} y_i \tag{6.25}$$

and

$$\sum_{i=1}^{N} z_{ij} y_i, \qquad j = 1, 2 \ldots, r. \tag{6.26}$$

The likelihood equations are found in the usual manner by setting the minimal sufficient statistics equal to their expected values:

$$\sum_{i=1}^{N} \hat{p}_i = \sum_{i=1}^{N} y_i, \tag{6.27}$$

$$\sum_{i=1}^{N} z_{ij} \hat{p}_i = \sum_{i=1}^{N} z_{ij} y_i, \qquad j = 1, 2, \ldots, r. \tag{6.28}$$

There are $r + 1$ likelihood equations for the $r + 1$ $\beta$s in model (6.24). The solution to equations (6.27) and (6.28) must be found numerically by means of some sort of iterative computational procedure. Haberman [1974a], for example, suggests the use of a modified Newton–Raphson procedure (see also the algorithms used by Gokhale [1972], Nelder and Wedderburn [1972], Nerlove and Press [1973], and Walker and Duncan [1967]). The modified Newton–Raphson algorithm has quadratic convergence properties, as opposed to the linear convergence properties of the iterative proportional fitting procedure used for all other problems in this book. For further details, including a detailed discussion of the large-sample theory for the general logistic regression model, see Haberman [1974a].

## 6.5 Polytomous and Multivariate Response Variables

Although the techniques considered in earlier sections have dealt only with a single dichotomous response variable, they may easily be extended to handle both (1) polytomous response variables and (2) two or more (possibly polytomous) response variables.

Suppose we have a single $I$-category response variable and two explanatory variables (with $J$ and $K$ categories, respectively). To describe the relationship between the response and explanatory variables we need a set of $I - 1$ logit models. But there are several ways we might choose to define the $I - 1$ logits. Four possibilities are:

$$\log\left(\frac{m_{ijk}}{\Sigma_{l>i} m_{ljk}}\right), \qquad i = 1, 2, \ldots, I - 1; \tag{6.29}$$

$$\log\left(\frac{m_{ijk}}{\Sigma_{l} m_{ljk}}\right), \qquad i = 1, 2, \ldots, I - 1; \tag{6.30}$$

$$\log\left(\frac{m_{ijk}}{m_{Ijk}}\right), \qquad i = 1, 2, \ldots, I - 1; \tag{6.31}$$

$$\log\left(\frac{m_{ijk}}{m_{i+1,jk}}\right), \qquad i = 1, 2, \ldots, I - 1. \tag{6.32}$$

When the response categories have a natural order, such as educational attainment (grade school, high school, college, graduate school), the choice of logits in (6.29) may be preferable. The quantities $m_{ijk}/\Sigma_{l>i} m_{ljk}$ are often referred to as *continuation ratios*, and they are of substantive interest in various fields. There is also a technical reason for working with the logarithms of the continuation ratios. Let $P_{ijk}$ be the probability of a response in category $i$ of variable 1 given levels $j$ and $k$ of the explanatory variables, where $\Sigma_i P_{ijk} = 1$. Then, when the $\{x_{ijk}\}$ are observed counts from $JK$ independent multinomials with sample sizes $\{x_{+jk}\}$ and cell probabilities $\{P_{ijk}\}$,

$$\frac{m_{ijk}}{\Sigma_{l \geqslant i} m_{ljk}} = \frac{P_{ijk}}{\Sigma_{l \geqslant i} P_{ijk}}. \tag{6.33}$$

We can write the multinomial likelihood functions as products of $I - 1$ binomial likelihoods, the $i$th of which has sample sizes $\{\Sigma_{l \geqslant i} x_{ljk}\}$ and cell probabilities $\{P_{ijk}/\Sigma_{l \geqslant i} P_{ljk}\}$. This means that if we use the method of maximum likelihood to estimate the parameters in the set of logit models, then we can do the estimation separately for each logit model, and we can simply add individual chi-square statistics to get an overall goodness-of-fit statistic for the set of models. Moreover, the observed binomial proportions,

$$\frac{x_{ijk}}{\Sigma_{l \geqslant i} x_{ljk}}, \qquad i = 1, 2, \ldots, I - 1, \tag{6.34}$$

are each asymptotically independent of the others so that we can assess the fit of the $I - 1$ logit models and various associated reduced models independently. Fienberg and Mason [1977] use this approach involving models for log continuation ratios in the context of logit models with simultaneous effects for age, period, and cohort. The interested reader is referred to that paper, and we shall not elaborate further here.

If we would like our logit models to correspond to loglinear models the summations in the denominators of (6.29) and (6.30) make these alternatives undesirable. Since we would like to fit similar models for all the logits (i.e., with the same effects involving the explanatory variables), working with the logits defined by (6.31) is equivalent to working with those defined by (6.32). For example, if $I = 3$, there are only three possible logits involving ratios of the form $m_{ijk}/m_{ljk}$ with $l > i$:

$$\log\left(\frac{m_{1jk}}{m_{2jk}}\right), \quad \log\left(\frac{m_{1jk}}{m_{3jk}}\right), \quad \log\left(\frac{m_{2jk}}{m_{3jk}}\right).$$

Clearly one of these is redundant since

$$\log\left(\frac{m_{1jk}}{m_{2jk}}\right) + \log\left(\frac{m_{2jk}}{m_{3jk}}\right) = \log\left(\frac{m_{1jk}}{m_{3jk}}\right). \tag{6.35}$$

Suppose we fit a loglinear model to the $I \times J \times K$ array that holds fixed the two-dimensional marginal total corresponding to the explanatory variables. Then the loglinear model can be transformed into a set of $I - 1$ nonredundant logit models of the form $\log(m_{ijk}/m_{1jk})$. Because different investigators might choose to look at different sets of logit models, it is reasonable to report the estimated interaction $u$-terms involving the effects of the explanatory variables on the response variable, rather than going to the trouble of constructing the estimated logit models.

In 1963 the President's Task Force on Manpower Conservation interviewed a national sample of 2400 young males rejected for military service because of failure to pass the Armed Forces Qualification Test. Hansen, Weisbrod, and Scanlon [1970] reanalyzed some of the data from this survey in an attempt to catalog the determinants of wage level for these low achievers. Table 6-4 offers a $3 \times 4 \times 2 \times 2$ four-dimensional cross-classification of 2294 armed forces rejectees analyzed by Talbot and Mason [1975]. The four variables are respondent's education (grammar school, some high school, high school graduate), father's education (grammar school, some high school, high school graduate, not available [NA]), age (less than 22, 22 or more), and race (white, black). For the present analysis we view the respondent's education as the response variable and the remaining three variables as

**Table 6-4**
Observed Cross-Classification of 2294 Young Males Who Failed to Pass the Armed Forces Qualification Test (Talbot and Mason [1975])

| Race | Age | Father's Education* | Respondent's Education: Grammar School | Some HS | HS Graduate |
|---|---|---|---|---|---|
| White | < 22 | 1 | 39 | 29 | 8 |
| | | 2 | 4 | 8 | 1 |
| | | 3 | 11 | 9 | 6 |
| | | 4 | 48 | 17 | 8 |
| | ≥ 22 | 1 | 231 | 115 | 51 |
| | | 2 | 17 | 21 | 13 |
| | | 3 | 18 | 28 | 45 |
| | | 4 | 197 | 111 | 35 |
| Black | < 22 | 1 | 19 | 40 | 19 |
| | | 2 | 5 | 17 | 7 |
| | | 3 | 2 | 14 | 3 |
| | | 4 | 49 | 79 | 24 |
| | ≥ 22 | 1 | 110 | 133 | 103 |
| | | 2 | 18 | 38 | 25 |
| | | 3 | 11 | 25 | 18 |
| | | 4 | 178 | 206 | 81 |

*1 = Grammar School, 2 = Some HS, 3 = HS Graduate, 4 = Not Available.

**Table 6-5**
Various Loglinear Models Fit to the 3 × 4 × 2 × 2 Cross-Classification in Table 6-4.

| Model | d.f. | $G^2$ |
|---|---|---|
| [234] [1] | 30 | 254.8 |
| [234] [12] | 24 | 162.6 |
| [234] [13] | 28 | 242.7 |
| [234] [14] | 28 | 152.8 |
| [234] [12] [13] | 22 | 151.5 |
| [234] [12] [14] | 22 | 46.7 |
| [234] [13] [14] | 26 | 142.5 |
| [234] [12] [13] [14] | 20 | 36.9 |
| [234] [123] [14] | 14 | 27.9 |
| [234] [124] [13] | 14 | 18.1 |
| [234] [134] [12] | 18 | 33.2 |
| [234] [123] [124] | 8 | 9.7 |

explanatory. Because there are so many missing observations under father's education, we have included "not available" as an extra category. Chen and Fienberg [1974] and Hocking and Oxspring [1974] describe alternative methods for dealing with such missing data.

Table 6-5 summarizes the fit of various loglinear models that treat the three-dimensional marginal totals for the explanatory variables as fixed. The simplest such model providing a good fit to the data is

$$[234]\ [124]\ [13]. \tag{6.29}$$

Four sets of estimated $u$-terms are needed in this model if we are to construct the corresponding logit models: $\{\hat{u}_{12(ij)}\}$, $\{\hat{u}_{13(ik)}\}$, $\{\hat{u}_{14(il)}\}$, $\{\hat{u}_{124(ijl)}\}$. There are three possible logit models corresponding to this loglinear model, and the estimated logit effects corresponding to these estimated $u$-terms are listed in Table 6-6.

The interpretation of the estimated effects listed in Table 6-6 is complex, and so we restrict ourselves here to an interpretation of the logit equation comparing "some high school" with "high school graduates." The estimated effect of age follows from the fact that older men have more time to complete high school. The first-order effects of race and father's education are best combined with the second-order effects. Then we see a declining effect of father's education for whites, and an increasing (but less dramatic) effect of father's education for blacks, on the log-odds of not completing versus completing high school. We cannot, however, ignore the effect of the missing observations for father's education. The size of the estimated parameter for

**Table 6-6**
Estimated Logit Effects for the Three Logit Models Corresponding to the Loglinear Model (6.29)

| | | Grammar vs. Some HS $\log(m_{1jkl}/m_{2jkl})$ | Grammar vs. HS Grad $\log(m_{1jkl}/m_{3jkl})$ | Some HS vs. HS Grad $\log(m_{2jkl}/m_{3jkl})$ |
|---|---|---|---|---|
| | Constant | −0.289 | 0.451 | 0.740 |
| Race | White | 0.395 | 0.390 | −0.005 |
| | Black | −0.395 | −0.390 | 0.005 |
| Age | < 22 | −0.120 | 0.099 | 0.219 |
| | ≥ 2 | 0.120 | −0.099 | −0.219 |
| Father's Education | Grammar S. | 0.380 | 0.406 | 0.026 |
| | Some HS | −0.371 | −0.355 | 0.016 |
| | HS Grad | −0.441 | −0.918 | −0.477 |
| | NA | 0.432 | 0.867 | 0.435 |
| White by | Grammar S. | 0.063 | 0.345 | 0.282 |
| | Some HS | −0.128 | −0.016 | 0.112 |
| | HS Grad | 0.030 | −0.429 | −0.459 |
| | NA | 0.035 | 0.101 | 0.066 |
| Black by | Grammar S | −0.063 | −0.345 | −0.282 |
| | Some HS | 0.128 | 0.016 | −0.112 |
| | HS Grad | −0.030 | 0.429 | 0.459 |
| | NA | −0.035 | −0.101 | −0.066 |

NA in the fitted model indicates that these data are not missing at random, and in fact the NA proportion differs somewhat by race.

Since loglinear models are well suited to handling multiple categorical response variables, no further analytical problems are introduced when we have categorical explanatory variables as well. If we condition on the values of the explanatory variables (as we do for logit models), then we can use loglinear models to assess the effects of the explanatory variables on the individual response variables and at the same time determine the interrelationships among the response variables.

# 7
# Causal Analysis Involving Logit and Loglinear Models

## 7.1 Path Diagrams

Many investigators are interested in providing a causal interpretation of the statistical relationships they find in the course of modeling a set of data. For quantitative variables the method of path analysis has been used to provide an interpretation of linear model systems (see, for example, Duncan [1966, 1975a] and Wright [1960]). Path analysis is not a method for discovering causal links among variables from the values of correlation coefficients. Rather, its role is

(1) to provide a causal interpretation for a given system of linear relationships,
(2) to make substantive assumptions regarding causal relationships explicit, thus avoiding internal inconsistencies.

With regard to point (1), we note that it is often the case that several causal models are consistent with a given set of relationships, and only additional information, substantive theory, or further research can help us choose among these models.

One of the salient features of path analysis as it is used in practice for the analysis of quantitative data is the diagrammatic representation of linear systems by means of arrows indicating various types of causal connections; then the "calculus" of path coefficients allows us to calculate numerical values for both direct and indirect effects, and these, in turn, are associated with the arrows in the path diagrams. Goodman [1972, 1973a, b] has recently proposed an analog to path analysis for qualitative variables, that is, for cross-classified data. The major feature of Goodman's approach is the creation of path diagrams, based on one or more loglinear or logit models and similar to those used in the analysis of quantitative variables. While Goodman does assign numerical values to the arrows in his diagrams, we note the following places where the analogy to the regular path analysis approach breaks down. For the analysis of categorical variables:

(i) there is no calculus of path coefficients;
(ii) because there is no calculus for path coefficients, there is no formal way to decide what values to assign to arrows not explicitly accounted for by a system of logit models, except by incomplete analogy with the regular path analysis techniques;
(iii) multiple categories for a variable lead to multiple coefficients to be associated with a given arrow in the diagram;
(iv) the existence of three-factor and higher-order interaction terms in a loglinear model lead to a complexity that cannot be handled without resorting to a more involved diagrammatic representation.

We do not view point (iv) as a serious obstacle, and for systems of binary variables, such as those Goodman analyzes, point (iii) presents no problem. Because of points (i) and (ii), however, we view the assignment of numerical values as problematic, and we would limit ourselves to an indication of sign for causal relationships, in a fashion similar to that described by Blalock [1964].

We begin by illustrating some basic ideas using a three-dimensional example. In a retrospective study of premarital contraceptive usage, Reiss, Banwart, and Foreman [1975] took samples of undergraduate female university students. One sample consisted of individuals who had attended the university contraceptive clinic, and the other was a control group consisting of females who had not done so. A preliminary analysis of data gathered indicated that the two samples did not differ significantly in terms of various background variables such as age, years in school, parents' education, etc. The individuals in the two samples were then cross-classified according to their virginity (virgin, nonvirgin) and various attitudinal variables such as belief that extramarital coitus is not always wrong. Table 7-1 displays the $2 \times 2 \times 2$ cross-classification corresponding to this particular attitudinal variable.

Although the totals corresponding to the use or nonuse of the clinic are fixed by design, we are actually interested in models that exhibit causal links from the other variables to the clinic variable. Figure 7-1 shows possible path diagrams that are consistent with such a causal framework. In addition, the two-dimensional table relating virginity and clinic use shows a strong relationship between the two variables.

In the diagrams of Figure 7-1, attitudinal response is denoted as E (variable 1), virginity as V (variable 2), and clinic use as C (variable 3). Curved double-headed arrows are used to denote relationships corresponding to nonzero

**Table 7-1**
Data on Premarital Contraceptive Usage From Reiss, Banwart, and Foreman [1975]

| | | Use of Clinic | | | |
|---|---|---|---|---|---|
| | | Yes | | No | |
| | | Virgin | Nonvirgin | Virgin | Nonvirgin |
| Attitude on Extramarital Coitus | Always Wrong | 23 | 127 | 23 | 18 |
| | Not Always Wrong | 29 | 112 | 67 | 15 |

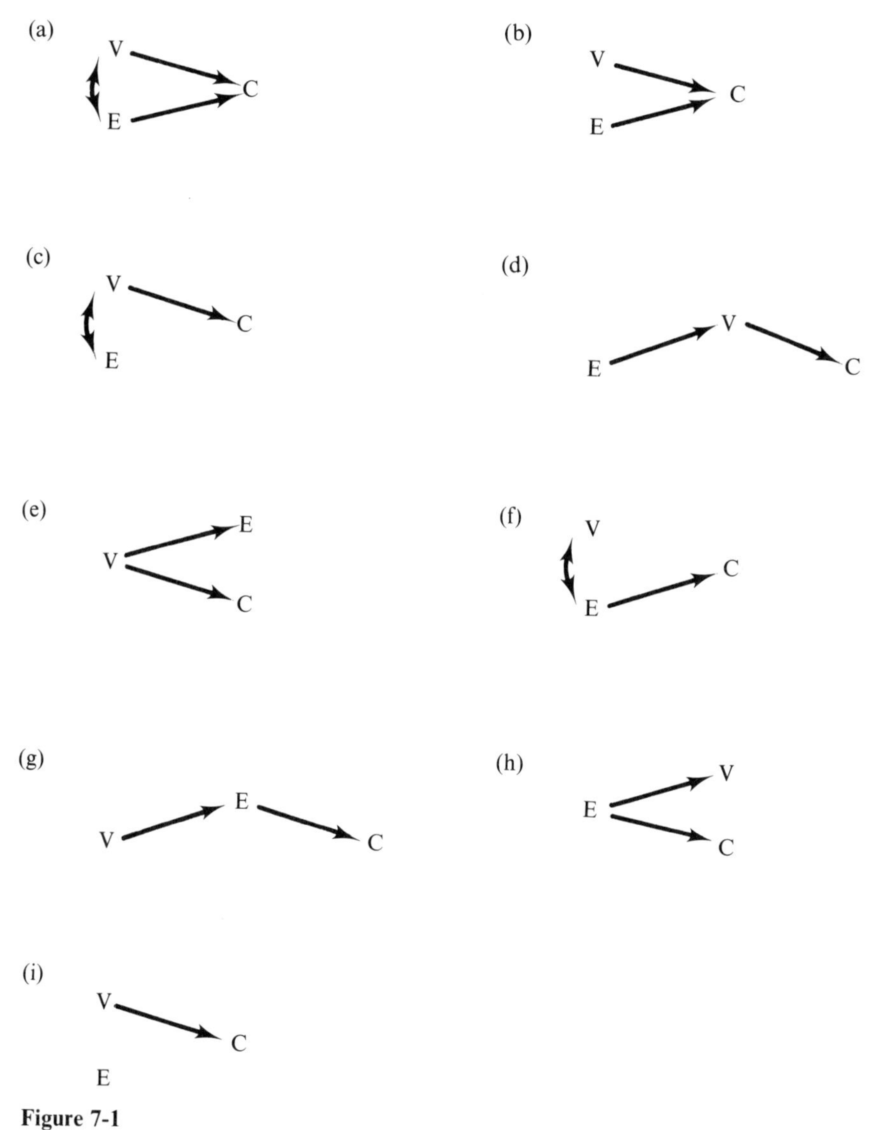


**Figure 7-1**
Some Path Diagrams Illustrating Possible Causal Connections among Variables for the Data in Table 7-1 (V = Virginity; E = Attitudinal Response; C = Clinic Attendance)

$u$-terms, which are not necessarily causal in nature. Figure 7-1a is consistent only with the no three-factor interaction model, $u_{123} = 0$. Figure 7-1b is consistent with $u_{12} = u_{123} = 0$; Figures 7-1c, d, and e are consistent with $u_{13} = u_{123} = 0$; and Figures 7-1f, g, and h are consistent with $u_{23} = u_{123} = 0$. Finally, Figure 7-1i is consistent with $u_{12} = u_{13} = u_{123} = 0$. This distinction among Figures 7-1c, d, and e comes from the causal relationships involving the explanatory variables. In Figure 7-1c, virginity and the attitudinal variable are not necessarily causally related; in Figure 7-1d, virginity is an *intervening* variable between attitude and clinic use; in Figure 7-1e, virginity is *antecedent* to both attitude and clinic use.

Analyses of the data in Table 7-1 will show that the only two unsaturated loglinear models providing an acceptable fit to the data are $u_{123} = 0$, and $u_{12} = u_{123} = 0$. Moreover, the no second-order interaction model does not provide a significantly better fit than the simpler model. The causal diagram consistent with the simpler model is given in Figure 7-1b.

## 7.2 Recursive Systems of Logit Models

When working with more than three categorical variables we must take care in carrying out analyses leading to the preparation of path diagrams. Implicit in the methods proposed by Goodman [1973a] is the causal ordering of variables. Suppose, for example, that we have four variables: $A$, $B$, $C$, and $D$. If the causal ordering is

$$A \text{ precedes } B \text{ precedes } C \text{ precedes } D, \tag{7.1}$$

then we should construct a diagram based on logit models for (1) $B$ given $A$, (2) $C$ given $A$ and $B$, and (3) $D$ given $A$, $B$, and $C$. This set of three logit models, when combined, characterizes the conditional joint probability of $B$, $C$, and $D$, given $A$. If the causal ordering is

$$\begin{pmatrix} A \\ \text{and} \\ B \end{pmatrix} \text{precede } C \text{ precedes } D, \tag{7.2}$$

then Goodman suggests that the relationship between $A$ and $B$ can be measured on the basis of the corresponding marginal table, and the links between the remaining variables can then be based on the logit models for $C$ given $A$ and $B$ and for $D$ given $A$, $B$, and $C$. This pair of logit models characterizes the conditional joint probability of $C$ and $D$ given $A$ and $B$. When we combine the two logit models with the marginal probability of $A$ and $B$

(jointly), we get a characterization of the joint probabilities associated with all four variables.

The key feature of expressions (7.1) and (7.2) is that they are both examples of what is referred to in the structural equations literature as a *recursive system of models*. That we should estimate the parameters in each of the logit equations in such a recursive system by the usual methods corresponds to the well-known results for recursive systems of linear regression models with normal error terms (see, for example, Wold and Jureen [1953]): the MLEs for the parameters in the system are just the MLEs of the parameters for each equation in the system when viewed separately.

We can assess the fit of a recursive system of logit models by directly checking the fit of each of the component models in the usual way, or by computing a set of estimated expected cell values for the combined system. The computation of these combined estimates is best illustrated by an example. Suppose we have four variables causally ordered as in expression (7.2). Then the estimated expected cell values for a system consisting of the pair of logit models implied by (7.2) are given by

$$\begin{aligned}\hat{m}^{*}_{ijkl} &= \frac{\hat{m}^{AB|C}_{ijk}\,\hat{m}^{ABC|D}_{ijkl}}{\hat{m}^{ABC|D}_{ijk+}},\\ &= \frac{\hat{m}^{AB|C}_{ijk}\,\hat{m}^{ABC|D}_{ijkl}}{x_{ijk+}},\end{aligned} \tag{7.3}$$

where $\{\hat{m}^{AB|C}_{ijk}\}$ are the estimated expected values for the logit model with variable $C$ as the response variable and $\{\hat{m}^{ABC|D}_{ijkl}\}$ are the estimated expected values for the logit model with variable $D$ as the response variable. Since the latter model involves conditioning on the marginal totals $\{x_{ijk+}\}$, we get the second equality in expression (7.3). Next suppose we have the causal ordering implied by expression (7.1). Then the estimated expected values for the system of three logit models implied by (7.1) are given by

$$\hat{m}^{**}_{ijkl} = \frac{\hat{m}^{A|B}_{ij}\,\hat{m}^{AB|C}_{ijk}\,\hat{m}^{ABC|D}_{ijkl}}{x_{ij++}\,x_{ijk+}}, \tag{7.4}$$

where $\{\hat{m}^{A|B}_{ij}\}$, $\{\hat{m}^{AB|C}_{ijk}\}$, and $\{\hat{m}^{ABC|D}_{ijkl}\}$ are the estimated expected values for the logit models with variables $B$, $C$, and $D$, respectively, as response variables.

Because of the simple multiplicative form of the estimated expected values in (7.3) and (7.4), the likelihood-ratio statistic $G^2$ for testing the fit of the combined system is simply the sum of the likelihood-ratio statistics for the component models. For example, for the system implied by (7.2) with the expected values given by expression (7.3), if

$$G_*^2 = 2 \sum_{ijkl} x_{ijkl} \log(x_{ijkl}/\hat{m}_{ijkl}^*), \tag{7.5}$$

$$G_{AB|C}^2 = 2 \sum_{ijk} x_{ijk+} \log(x_{ijk+}/\hat{m}_{ijk}^{AB|C}), \tag{7.6}$$

and

$$G_{ABC|D}^2 = 2 \sum_{ijkl} x_{ijkl} \log(x_{ijkl}/\hat{m}_{ijkl}^{ABC|D}), \tag{7.7}$$

then

$$G_*^2 = G_{AB|C}^2 + G_{ABC|D}^2. \tag{7.8}$$

Similarly, for the system implied by (7.1) with expected values given by expression (7.4), if

$$G_{**}^2 = 2 \sum_{ijkl} x_{ijkl} \log(x_{ijkl}/\hat{m}_{ijkl}^{**}) \tag{7.9}$$

and

$$G_{A|B}^2 = 2 \sum_{ij} x_{ij++} \log(x_{ij++}/\hat{m}_{ij}^{A|B}), \tag{7.10}$$

then

$$G_{**}^2 = G_{A|B}^2 + G_{AB|C}^2 + G_{ABC|D}^2. \tag{7.11}$$

Goodman [1973a, b] has used these techniques to analyze an example presented by Coleman [1964]. Each of 3398 schoolboys was interviewed at two points in time and asked about his self-perceived membership in the "leading crowd" (in = +; out = −) and his attitude toward the "leading crowd" (favorable = +; unfavorable = −). The resulting data are reproduced in Table 7-2. This example is an illustration of what is commonly referred to as a *two-wave, two-variable panel model*. In order to continue with the notation used earlier in this section, we label membership and attitude in the first interview by $A$ and $B$, respectively, and in the second interview by $C$ and $D$, respectively.

What assumptions regarding the causal ordering of variables might one assume for data in this example? One plausible assumption is that $A$ and $B$ precede $C$ and $D$ (they clearly do so in a temporal sense). A second, more speculative assumption might be that attitudes toward the "leading crowd"

**Table 7-2**
Two-Wave Two-Variable Panel Data for 3398 Schoolboys: Membership in and Attitude toward the "Leading Crowd" (Coleman [1964])

| | | | Second Interview | | | |
|---|---|---|---|---|---|---|
| | | Membership | + | + | − | − |
| | | Attitude | + | − | + | − |
| | Member-ship | Attitude | | | | |
| First Interview | + | + | 458 | 140 | 110 | 49 |
| | + | − | 171 | 182 | 56 | 87 |
| | − | + | 184 | 75 | 531 | 281 |
| | − | − | 85 | 97 | 338 | 554 |

affect membership in it, which would result in the causal order suggested by expression (7.1). The relationship between variables $A$ and $B$ is measured by the cross-product ratio ($\hat{\alpha} = 1.53$) of the two-dimensional marginal table or $\frac{1}{4} \log \hat{\alpha}$ (see expression (2.22)):

| | | Attitude ($B$) + | − |
|---|---|---|---|
| Membership ($A$) | + | 757 | 496 |
| | − | 1071 | 1074 |

To test whether $\alpha = 1$, we compute the likelihood-ratio statistic: $G^2_{A|B} = 35.1$, a value in the extreme right-hand tail of the $\chi^2_1$ distribution. Thus we conclude $A$ and $B$ are positively related. Note that we are not in a position, on the basis of the data themselves, to distinguish between the causal orderings "$A$ precedes $B$" and "$B$ precedes $A$" because of the symmetry inherent in the form of the cross-product ratio. Because $A$ and $B$ are assumed to be related in our model building, we have $\hat{m}_{ij}^{A|B} = x_{ij++}$, and so the combined estimates in expression (7.4) reduce in form to those described by expression (7.3).

Next we turn to building a logit model with $C$ (membership at the time of the second interview) as the response variable and $A$ and $B$ as explanatory. The three unsaturated loglinear models corresponding to such a logit model are

1. $[AB][AC][BC]$ (1 d.f.; $G^2 = 0.0$)
2. $[AB][BC]$ (2 d.f.; $G^2 = 1005.1$)
3. $[AB][AC]$ (2 d.f.; $G^2 = 27.2$).

The only one of these models that provides an acceptable fit to the data is model 1 (no second-order interaction). Model 1 corresponds to the logit model

$$\text{logit}_{ij}^{AB|C} = \log\left(\frac{m_{ij1}^{AB|C}}{m_{ij2}^{AB|C}}\right) = w^{AB|C} + w_{1(i)}^{AB|C} + w_{2(j)}^{AB|C}, \tag{7.12}$$

and the estimated effect parameters for this model are $\hat{w}_{1(1)}^{AB|C} = 1.24$ and $\hat{w}_{2(1)}^{AB|C} = 0.22$.

Finally, we turn to building a logit model with $D$ (attitude at the time of the second interview) as the response variable and $A$, $B$, and $C$ as explanatory. There are eight unsaturated loglinear models corresponding to such a logit model with no second-order interaction effects (each of three two-factor effects may be present or absent), and we list four of these here:

4. $[ABC][AD][BD][CD]$ (4 d.f.; $G^2 = 1.2$)
5. $[ABC][BD][CD]$ (5 d.f.; $G^2 = 4.0$)
6. $[ABC][AD][CD]$ (5 d.f.; $G^2 = 262.5$)
7. $[ABC][AD][BD]$ (5 d.f.; $G^2 = 15.7$).

Models 4 and 5 both provide an acceptable fit to the data, while models 6 and 7 do not. Since model 5 is a special case of model 4, and since the difference $G^2(5) - G^2(4) = 4.0 - 1.2 = 2.8$ is not significant when compared with the 0.05 tail value for the $\chi_1^2$ distribution, our preferred model is model 5. The corresponding logit model is

$$\text{logit}_{ijk}^{ABC|D} = w^{ABC|D} + w_{2(j)}^{ABC|D} + w_{3(k)}^{ABC|D}, \tag{7.13}$$

and the estimated effect parameters for this model are $\hat{w}_{2(1)}^{ABC|D} = 0.58$ and $\hat{w}_{3(1)}^{ABC|D} = 0.21$. The estimated expected cell values (computed using expression (7.3)) for the recursive system of logit models are displayed in Table 7-3, and the system is summarized by the diagram in Figure 7-2. Note that, following

**Table 7-3**
Expected Values for System of Logit Models Given by Expressions (7.12) and (7.13)

| | | | Second Interview | | | |
|---|---|---|---|---|---|---|
| | | Membership | + | + | − | − |
| | | Attitude | + | − | + | − |
| | Membership | Attitude | | | | |
| | + | + | 447.7 | 151.2 | 104.6 | 53.5 |
| First | + | − | 169.4 | 182.8 | 54.6 | 89.3 |
| Interview | − | + | 193.0 | 65.2 | 537.7 | 275.2 |
| | − | − | 88.0 | 94.9 | 338.1 | 553.0 |

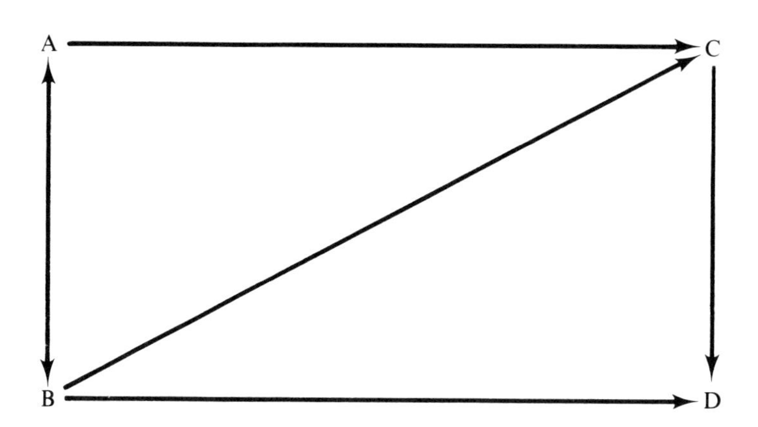


First Interview Second Interview

**Figure 7-2**
Path Diagram Showing Causal Connections Implied by the Logit Models (7.12) and (7.13)

our earlier discussion, we have not assigned any numerical values to the arrows. Also, the arrow going between $A$ and $B$ has two heads since we cannot distinguish between $A$ causing $B$ and $B$ causing $A$. For the combined recursive system we have $G_*^2 = 4.1$ with 6 d.f., clearly a good fit. The difference between $G_*^2$ and $G_{AB|C}^2 + G_{ABC|D}^2 = 4.0$ is due to rounding error.

The causal interpretation of Figure 7-2 is that (1) membership and attitude at the time of the first interview are related; (2) membership at the time of the second interview is affected by both membership and attitude at the time of the first interview (and there is no second-order effect of these two variables), as indicated by equation (7.12); and (3) attitude in the second interview is affected by concurrent membership and previous attitude, as indicated by equation (7.13).

There are other recursive systems of logit models, based on other assumptions, which provide just as good a fit to these data as the system in Figure 7-2. For example, Goodman [1973b] describes a system with an apparent time reversal based on markedly different causal assumptions.

If we were not especially concerned about structuring the variables in this example in terms of a series of logit models, Figure 7-2 might suggest that we consider fitting the following loglinear model to the four-dimensional array, with a two-factor effect corresponding to each arrow in the diagram:

8. $[AB]\ [AC]\ [BC]\ [BD]\ [CD]$.

The likelihood-ratio goodness-of-fit statistic for this model has a value $G^2 = 4.1$ with 6 d.f. It is not a coincidence that this numerical value and the associated degrees of freedom are identical to those for $G^2_*$, that is, for the recursive pair of logit models. Given that model 5 fits the data, the generalization of Theorem 3-1 on collapsing tables (p. 45) says that we can test for the absence of the second-order interaction involving variables $A$, $B$, and $C$ in the marginal table involving those variables by fitting model 1. It then follows from the standard results on partitioning $G^2$ that the sum of the likelihood-ratio statistics for models 1 and 5 equals the likelihood-ratio statistic for model 8. *We note that fitting recursive systems of logit models is not always equivalent to fitting a single loglinear model to the full cross-classification.*

## 7.3 Recursive Systems: A More Complex Example

While the example in the preceding section illustrates the basic features of the analysis of recursive systems of logit models, it does not illustrate all of the difficulties alluded to in Section 7.1. The following example includes polytomous variables and logit models with second-order interaction effects.

In a study of a randomly selected cohort of 10,318 Wisconsin high school seniors, Sewell and Shah [1968] explored the relationship among five variables: ($A$) socioeconomic status (high, upper middle, lower middle, low), ($B$) intelligence (high, upper middle, lower middle, low) as measured by the Hemmon–Nelson Test of Mental Ability, ($C$) sex (male, female), ($D$) parental encouragement (low, high), and ($E$) college plans (yes, no). The five-dimensional cross-classification is reproduced here as Table 7-4. In the course of their analysis, Sewell and Shah propose a causal model that can be described by

$$\begin{pmatrix} A \\ \text{and} \\ B \\ \text{and} \\ C \end{pmatrix} \text{precede } D \text{ precedes } E. \tag{7.14}$$

Corresponding to this causal structure is a pair of logit models: one with parental encouragement as a response variable and socioeconomic status (SES), intelligence (IQ), and sex as explanatory, and the other with college plans as the response variable and all four remaining variables as explanatory.

We develop the two logit models separately, beginning with the first one with variable $D$, parental encouragement, as the response variable. Some plausible unsaturated logit models are:

**Table 7-4**
Social Class, Parental Encouragement, IQ, and Educational Aspirations (Sewell and Shah, 1968)

| Sex | IQ | College Plans | Parental Encouragement | SES L | SES LM | SES UM | SES H |
|---|---|---|---|---|---|---|---|
| M | L | Yes | Low | 4 | 2 | 8 | 4 |
| | | | High | 13 | 27 | 47 | 39 |
| | | No | Low | 349 | 232 | 166 | 48 |
| | | | High | 64 | 84 | 91 | 57 |
| | LM | Yes | Low | 9 | 7 | 6 | 5 |
| | | | High | 33 | 64 | 74 | 123 |
| | | No | Low | 207 | 201 | 120 | 47 |
| | | | High | 72 | 95 | 110 | 90 |
| | UM | Yes | Low | 12 | 12 | 17 | 9 |
| | | | High | 38 | 93 | 148 | 224 |
| | | No | Low | 126 | 115 | 92 | 41 |
| | | | High | 54 | 92 | 100 | 65 |
| | H | Yes | Low | 10 | 17 | 6 | 8 |
| | | | High | 49 | 119 | 198 | 414 |
| | | No | Low | 67 | 79 | 42 | 17 |
| | | | High | 43 | 59 | 73 | 54 |
| F | L | Yes | Low | 5 | 11 | 7 | 6 |
| | | | High | 9 | 29 | 36 | 36 |
| | | No | Low | 454 | 285 | 163 | 50 |
| | | | High | 44 | 61 | 72 | 58 |
| | LM | Yes | Low | 5 | 19 | 13 | 5 |
| | | | High | 14 | 47 | 75 | 110 |
| | | No | Low | 312 | 236 | 193 | 70 |
| | | | High | 47 | 88 | 90 | 76 |
| | UM | Yes | Low | 8 | 12 | 12 | 12 |
| | | | High | 20 | 62 | 91 | 230 |
| | | No | Low | 216 | 164 | 174 | 48 |
| | | | High | 35 | 85 | 100 | 81 |
| | H | Yes | Low | 13 | 15 | 20 | 13 |
| | | | High | 28 | 72 | 142 | 360 |
| | | No | Low | 96 | 113 | 81 | 49 |
| | | | High | 24 | 50 | 77 | 98 |

1. [ABC][AD][BD][CD] (24 d.f.; $G^2 = 55.81$)
2. [ABC][ABD][CD] (15 d.f.; $G^2 = 34.60$)
3. [ABC][BCD][ACD] (18 d.f.; $G^2 = 31.48$)
4. [ABC][ABD][BCD] (12 d.f.; $G^2 = 22.44$)
5. [ABC][ABD][ACD] (12 d.f.; $G^2 = 22.45$)
6. [ABC][ABD][ACD][BCD] ( 9 d.f.; $G^2 = 9.22$).

The fit of model 1, with only first-order logit effects of the explanatory variables, is significant at the 0.001 level, and an examination of the residuals strongly suggests the inclusion of all three possible second-order effects. Models 2, 3, 4, and 5 include one or two such effects, and while they represent significant improvements (at the 0.01 level) over model 1, the fit of each is still

**Table 7-5**
Estimated Logit Effect Parameters For Model 6, Predicting Parental Encouragement, Fitted to Data from Table 7-4

(i) $w^{ABC|D} = -0.124$

(ii) $\{w_{1(i)}^{ABC|D}\} = \{1.178, 0.384, -0.222, -1.340\}$ (L, LM, UM, H)

(iii) $\{w_{2(j)}^{ABC|D}\} = \{0.772, 0.226, -0.210, -0.788\}$ (L, LM, UM, H)

(iv) $\{w_{3(k)}^{ABC|D}\} = \{-0.304, +0.304\}$ (M, F)

(v) $\{w_{1\,2(ij)}^{ABC|D}\}$ =

| (SES) \ (IQ) | L | LM | UM | H |
|---|---|---|---|---|
| L | 0.016 | −0.098 | 0.058 | 0.026 |
| LM | −0.066 | −0.032 | −0.144 | 0.244 |
| UM | −0.074 | 0.044 | 0.138 | −0.108 |
| H | 0.126 | 0.086 | −0.048 | −0.164 |

(vi) $\{w_{1\,3(ik)}^{ABC|D}\}$ =

| (SES) \ (Sex) | M | F |
|---|---|---|
| L | −0.140 | 0.140 |
| LM | 0.052 | −0.052 |
| UM | −0.018 | 0.018 |
| H | 0.106 | −0.106 |

(vii) $\{w_{2\,3(jk)}^{ABC|D}\}$ =

| (IQ) \ (Sex) | M | F |
|---|---|---|
| L | 0.126 | −0.126 |
| LM | 0.016 | −0.016 |
| UM | −0.018 | 0.018 |
| H | −0.122 | 0.122 |

significant at the 0.05 level. Finally model 6, with all three second-order effects, provides a very good fit to the data, and the differences between this model and each of models 3, 4, and 5 are all significant at the 0.01 level.

The estimated logit parameters for model 6 are given in Table 7-5. (All of the final digits are multiples of two since appropriate loglinear estimated effects were computed to three decimal places and then doubled.) Note that the first-order estimated effects of SES, IQ, and sex are as we might expect. The logit for low versus high parental encouragement decreases monotonically as we move from low to high SES and from low to high IQ, and males get more encouragement than females. The sex by IQ interactive (second-order) effects are monotonically ordered, with low IQ males receiving relatively less encouragement than low IQ females and high IQ males receiving relatively more encouragement than high IQ females. The other two sets of second-order effects are not monotonically ordered and are thus more difficult to interpret. The magnitude of second-order effects rarely exceeds two-thirds the magnitude of the smallest first-order effects.

Next we turn to the logit model for predicting college plans. Some plausible unsaturated models are:

7. $[ABCD][E]$ (63 d.f.; $G^2 = 4497.51$)
8. $[ABCD][AE][BE][CE][DE]$ (55 d.f.; $G^2 = 73.82$)
9. $[ABCD][BCE][AE][DE]$ (52 d.f.; $G^2 = 59.55$)

**Table 7-6**
Estimated Logit Effect Parameters for Model 9, Predicting College Plans, as Fitted to Data in Table 7-4

(i) $w^{ABCD|E} = -1.292$

(ii) $\{w_{1(i)}^{ABCD|E}\} = \{-0.650 \quad -0.200 \quad 0.062 \quad 0.790\}$ (L, LM, UM, H)

(iii) $\{w_{2(j)}^{ABCD|E}\} = \{-0.840 \quad -0.300 \quad 0.266 \quad 0.876\}$ (L, LM, UM, H)

(iv) $\{w_{3(k)}^{ABCD|E}\} = \{0.082 \quad -0.082\}$ (M, F)

(v) $\{w_{4(l)}^{ABCD|E}\} = \{-1.214 \quad 1.214\}$ (L, H)

(vi) $\{w_{23(jk)}^{ABCD|E}\}$ =

| | M | F |
|---|---|---|
| L | −0.134 | 0.134 |
| LM | −0.078 | 0.078 |
| UM | 0.094 | −0.094 |
| H | 0.118 | −0.118 |

(IQ)

10. $[ABCD][BCE][ACE][DE]$ (49 d.f.; $G^2 = 57.99$).

The fit of model 8, with only first-order logit effects, is barely significant at the 0.05 level, whereas the fit of model 9 corresponds to a 0.22 descriptive level of significance, and the difference in fit between the two is significant at the 0.005 level. Since model 10 does not give a significant improvement in fit over model 9, we conclude that model 9 is the most parsimonious model providing a good fit to the data.

Table 7-6 gives the estimated logit effect parameters for model 9. As in the Table 7-5, the first-order logit effects are as we might expect, with monotonically increasing effects of SES and IQ on college plans and positive effects associated with high parental encouragement and with being male. The second-order effects involving sex and IQ are again monotonic and show that a male with a low IQ is much less likely to plan for college than a female with a comparable IQ, whereas the opposite is true for males and females with high IQs.

The existence of second-order effects in the pair of logit equations, and the multiple categories for SES and IQ, make a simple diagrammatic representation of the logit system for the data in Table 7-5 impossible.

## 7.4 Nonrecursive Systems of Logit Models

The results on recursive systems of logit models described in Section 7.2 mimic the results for recursive systems of linear equations with additive error terms. In such systems the "causal linkages" all run in one direction.

Econometricians, and more recently sociologists and other social scientists, have spent considerable effort studying nonrecursive systems of linear equations with additive error terms. In these nonrecursive systems, the causal linkages between at least two response variables run both ways. For example, in a system involving four variables $X_1$, $X_2$, $X_3$, and $X_4$, the following pair of "structural" equations illustrate a nonrecursive system:

$$X_3 = \beta_{31}X_1 + \beta_{34}X_4 + \varepsilon_3, \tag{7.15}$$

$$X_4 = \beta_{42}X_2 + \beta_{43}X_3 + \varepsilon_4, \tag{7.16}$$

where $\varepsilon_1$ and $\varepsilon_2$ are random-error terms with zero means. Note how $X_4$ is used for predicting $X_3$ in equation (7.15), while $X_3$ is used for predicting $X_4$ in equation (7.16). The aim of this nonrecursive system is to explain the relationship between the jointly dependent variables, $X_3$ and $X_4$, in terms of the separate "causal effects" of $X_4$ on $X_3$ and of $X_3$ on $X_4$, with $\beta_{34}$ not necessarily equal to $1/\beta_{43}$. If we make appropriate assumptions for such systems, then we can identify or separate out these reciprocal effects. Duncan [1975a] provides an excellent introduction to both recursive and nonrecursive systems of linear structural equations with additive error terms, and econometrics textbooks such as Fisher [1966] and Goldberger [1964] give additional technical details.

Can we set up nonrecursive systems of logit models for categorical variables, with properties resembling those of the nonrecursive systems of linear structural equations? The answer to this question is unclear. We can best see the problems in the categorical case in the context of an example involving four dichotomous categorical variables, $A$, $B$, $C$, and $D$, with $A$ and $B$ occurring a priori to $C$ and $D$. We would like to predict simultaneously variations in $D$ using $B$ and $C$ and variations in $C$ using $A$ and $D$. One possible way to model such a situation is through the following pair of simultaneous logit models:

$$\log \frac{m_{ijk1}^{ABC|D}}{m_{ijk2}^{ABC|D}} = W^{ABC|D} + W_{2(j)}^{ABC|D} + W_{3(k)}^{ABC|D}, \tag{7.17}$$

$$\log \frac{m_{ij1l}^{ABD|C}}{m_{ij2l}^{ABD|C}} = W^{ABD|C} + W_{1(i)}^{ABD|C} + W_{4(l)}^{ABD|C}. \tag{7.18}$$

Variables $C$ and $D$ are both response and explanatory variables in this system. Equations (7.17) and (7.18) mimic the form of the linear structural equations (7.15) and (7.16). The parallel, however, is far from complete, and no one has yet proposed a method for consistently estimating the parameters in (7.17) and (7.18) so that the effect of $C$ on $D$ is not necessarily the same as the effect of $D$ on $C$. Fitting each of the logit models independently to a four-dimensional table produces unequal reciprocal effects of the response variables (see, for example, Nerlove and Press [1973]) but seems to ignore the simultaneity of the models. The only alternative approach proposed to date is to combine (7.17) and (7.18) into a single logit model that treats $C$ and $D$ as jointly dependent variables. This preserves the simultaneity but forces the reciprocal effects to be equal to one another (see, for example, Goodman [1973a] and Nerlove and Press [1973]).

### 7.5 Retrospective Epidemiological Studies

There is a specific class of studies, most common in epidemiological research, in which, instead of working forward (in a prospective manner) from explanatory variables to responses, investigators work in reverse. They start with the existence of a disease and work backwards in time to discover differences between those who have contracted the disease (i.e., cases) and those who have not (i.e., controls). In such retrospective studies we first identify the cases, then choose the controls, and finally compare the cases and controls to see how they differ. Thus the totals for the natural response variable—the presence or absence of disease—are fixed rather than the totals for one or more of the explanatory variables. Even though we appear to have done everything backwards, we can still estimate the effect of variable explanatory variables on the response variable as a result of properties associated with the cross-product ratio (and thus $u$-terms in loglinear models). This point is most easily illustrated in the context of a $2 \times 2$ table.

Let $D$ and $\bar{D}$ refer to the presence and absence of the disease, respectively, and $A$ and $\bar{A}$ to the presence and absence of some (potentially causal) characteristic. In the prospective study we fix the proportions associated with $A$ and $\bar{A}$ and then observe the numbers corresponding to $D$ and $\bar{D}$. The standard measure of the association between the two variables is the odds ratio,

$$\alpha_{PRO} = \frac{\Pr(D|A)\ \Pr(\bar{D}|\bar{A})}{\Pr(\bar{D}|A)\ \Pr(D|\bar{A})}. \tag{7.19}$$

In the retrospective study we fix the proportions for $D$ and $\bar{D}$ and then go back and find the numbers corresponding to $A$ and $\bar{A}$. The odds ratio relevant to this situation is

$$\alpha_{RET} = \frac{\Pr(A|D)\ \Pr(\bar{A}|\bar{D})}{\Pr(\bar{A}|D)\ \Pr(A|\bar{D})}. \tag{7.20}$$

By invoking either Bayes's theorem or the invariance property of the cross-product ratio discussed in Section 2.5, we immediately see that

$$\alpha_{PRO} = \alpha_{RET} \tag{7.21}$$

and thus the retrospective study measures the same relationship between the explanatory and response variables as does the prospective study.

When we do a retrospective study with several explanatory variables, we often focus on one of them as a potential causal agent and simply control for the others. A standard approach in such circumstances is to fit a logit model using the explanatory variable that is the potential causal agent as if it were the response, while placing conditions on all of the remaining variables. We illustrate this approach by an example.

Cochran [1954] presents an example comparing the mothers of Baltimore school children referred by their teachers as presenting behavioral problems with mothers of a comparable group of control children. The basic data are

**Table 7-7**
Data on Number of Mothers with Previous Infant Losses (Cochran [1954])

| | Birth Order | | | | | | Totals |
|---|---|---|---|---|---|---|---|
| | 2 | | 3–4 | | 5+ | | |
| Losses | Yes | No | Yes | No | Yes | No | |
| Problems | 20 | 82 | 26 | 41 | 27 | 22 | 218 |
| Controls | 10 | 54 | 16 | 30 | 14 | 23 | 147 |
| Totals | 30 | 136 | 42 | 71 | 41 | 45 | 365 |

reproduced in Table 7-7. The three variables involved in this study are (1) group (problems vs. controls), (2) infant losses (e.g., stillbirths) previous to the child in the study (yes, no), (3) birth order of child (2, 3–4, 5+). We are interested in the effects of previous losses and birth order on the outcome variable (problems vs. controls). Note that the number of problems and the number of controls are fixed by design.

Various authors (e.g., Grizzle [1961] and Kullback [1973]) have fitted logit models to these data, treating infant losses as the response variable and the other two variables as explanatory. The simple model of logit additivity (i.e., no second-order interaction) fits the data extremely well: $G^2 = 0.85$ with 3 d.f. The logit models that exclude the effect of problem vs. control and of birth order have associated goodness-of-fit values of $G^2 = 3.15$ and $G^2 = 28.20$ with 3 and 4 d.f., respectively. Thus birth order is related to previous infant losses, but problem vs. control seems to be unrelated. Finally, we turn the model around and note that there appears to be no difference between problem and control children as a result of previous losses.

Recently economists have proposed the use of retrospective studies of the case-control form under the label of "endogenous" or "choice-based" sampling (see, for example, Manski and Lerman [1976], Manski and McFadden [1977], and McFadden [1976]). Their logit models have essentially the same structure as the epidemiological models.

# 8
# Fixed and Random Zeros

Zero entries in contingency tables are of two types: *fixed* and *sampling* zeros. Fixed zeros occur when it is impossible to observe values for certain combinations of the variables (e.g., in a hospital study, we cannot find "male obstetrical patients," so that the zero is due to an honest zero probability for that cell). In Sections 8.2 and 8.3 we shall briefly indicate how to extend the models and methods already discussed to such *incomplete* tables.

Sampling zeros are due to sampling variation and the relatively small size of the sample when compared with the large number of cells; they disappear when we increase the sample size sufficiently (e.g., in a national study involving religious affiliation and occupation, we are not likely to find many Jewish farmers from Iowa, yet such individuals do exist).

## 8.1 Sampling Zeros and MLEs in Loglinear Models

For multidimensional contingency tables, in which the number of cells is often quite large, it is difficult to get a sufficiently large sample to rid ourselves of all the sampling zeros. One of the most powerful properties of the models and methods of estimation discussed so far is that cells with zero entries due to sampling variation can have nonzero expected values. For example, the estimated expected values for all of the models fit to the data in Table 3-6 are nonzero.

In order that the computed expected cell values for a particular model be all positive, we must make certain that, if there are two or more sampling zeros, when one observed zero cell is made positive, one of the remaining observed zero cells is not made negative. For example, consider a $2 \times 2 \times 2$ table with positive observed values except for the (1, 1, 1) and (2, 2, 2) cells (see the example in Table 8-1a). If the model being fitted is one with no three-factor interaction but all two-factor interactions, making the (1, 1, 1) cell positive automatically makes the (2, 2, 2) cell negative because the expected cell values must satisfy three sets of two-dimensional marginal constraints. It is often quite difficult to ascertain when situations such as the one in this example occur; however, when the iterative proportional fitting procedure is used to get expected values for models in tables with "too many" sampling zeros, it converges to a table with zero entries in some of the cells that originally had sampling zeros. If we examine the expected values after each cycle of the iteration, we can often detect that the values in some cells are converging to zero.

On the other hand, the computed expected values are always positive if there is only one sampling zero in a table. For example, suppose the (1, 1, 1) cell in a $2 \times 2 \times 2$ table has a zero observed value due to sampling variation

**Table 8-1**

(a) Example of a $2 \times 2 \times 2$ table for which there do not exist positive MLEs of the expected cell values under the model $u_{123} = 0$.

| | | | | |
|---|---|---|---|---|
| 0 | 5 | | 6 | 9 |
| 16 | 7 | | 5 | 0 |

(b) Example of a $2 \times 2 \times 2$ table with an observed two-dimensional marginal total

| | | | | |
|---|---|---|---|---|
| 0 | 5 | | 6 | 9 |
| 0 | 16 | | 5 | 7 |

while the remaining seven cells have nonzero observed values. Then each one-way and each two-dimensional observed marginal total is positive, and the expected value for the (1, 1, 1) cell will be positive under any of the seven models that do not include the three-factor interaction term.

The iterative estimation procedure presented here can be used with only one minor change when the observed marginal totals, required for a given loglinear model, have some zero entries. That change is to define zero divided by zero as zero. Thus, if an entry in an observed marginal total used for a given model is zero, all entries adding up to that total will necessarily remain equal to zero during the iterative procedure.

In order to test the goodness-of-fit of a model that uses an observed set of marginal totals with at least one zero entry, we must reduce the degrees of freedom associated with the test statistic. The reason for this is quite simple. If an observed marginal entry is zero, both the expected and the observed entries for all cells included in that total must be zero, and so the fit of the model for those cells is known to be perfect once it is observed that the marginal entry is zero. As a result, we must delete those degrees of freedom associated with the fit of the zero cell values.

For an example of how to reduce the degrees of freedom, consider a $2 \times 2 \times 2$ table with positive observed values except for the (1, 1, 1) and (2, 1, 1) cells (see Table 8-1b). Here the observed marginal total $x_{+11} = 0$. Now suppose that the totals $x_{+ij}$ are being used to get cell estimates for a given model. If we have the total $x_{+11}$, knowledge of $x_{111}$ is sufficient to determine $x_{211}$, and conversely. Thus there is one degree of freedom associated with the cells adding up to $x_{+11}$, and it must be subtracted from the total degrees of freedom when $x_{+11} = 0$. On the other hand, if the model being fitted has no three-factor interaction but all two-factor interactions, the total degrees of freedom is only one. If $x_{+11} = 0$ in that situation, there are no degrees of freedom left, and the fit of the model is perfect. The procedure for handling zero marginal totals discussed here would also be used if all the sampling zeros, corresponding to the observed zero marginal totals, were actually fixed zeros (see Sections 8.2 and 8.3).

A general formula for computing the degrees of freedom in cases where some of the margins fitted contain sampling zeros is as follows:

$$\text{d.f.} = (T_e - Z_e) - (T_p - Z_p), \tag{8.1}$$

where

$T_e$ = # cells in table that are being fitted,
$T_p$ = # parameters fitted by model,
$Z_e$ = # cells containing zero estimated expected values,
$Z_p$ = # parameters that cannot be estimated because of zero marginal totals.

Note how (8.1) generalizes our earlier formula (3.42) for degrees of freedom. The application of this formula to the two examples above is straightforward, but we must take care in more complicated cases where a zero cell entry corresponds to multiple zero marginal totals (see Bishop, Fienberg, and Holland [1975], pp. 114–119, for further details).

It is rather difficult to find contingency tables in the biological or social science literature that contain zero cell values, let alone zero marginal totals. This is mainly due to suggestions on the collapsing of categories that are found in most elementary statistical textbooks. It is my opinion, however, that when the categories for a given variable are truly meaningful, collapsing of categories is not necessarily a good procedure, especially given the availability of the methods just described.

## 8.2 Incomplete Two-Dimensional Contingency Tables

Many authors have concerned themselves with the analysis of contingency tables whose entries are missing, a priori zero, or otherwise predetermined. One way to treat such data is to remove these special cells from the model building and analysis process, after which we are left with a contingency table that is said to be structurally *incomplete*. Note that this type of incompleteness is akin to truncation of the full table, and it should not be confused with the incompleteness that results from not being able to completely cross-classify all of the individuals in a study. For a detailed discussion of the latter class of problems, see Chen and Fienberg [1974, 1976].

One of the most serious problems encountered in the application of incomplete contingency table methodology is that investigators who have data in the form of an incomplete contingency table often fail to recognize that fact. They either fill in the cells containing fixed zeros using some "appropriate values," or they collapse the data (i.e., they collapse either the number of categories for certain variables or the number of variables) until the fixed zeros have "vanished." Some times the occurrence of fixed zeros or missing entries leads the researcher to abandon analysis of the data. These practices can lead to inappropriate conclusions, general confusion, and bad science.

In this section we briefly illustrate the application to incomplete situations of the methodology for complete cross-classifications. There are several technical problems that we shall ignore here. Readers interested in more detailed accounts should refer to Bishop, Fienberg, and Holland [1975], Goodman [1968], and Haberman [1974a].

We begin with two-dimensional tables. Let $S$ be the set of cells in an $I \times J$ array that remain after the exclusion of missing entries and fixed values, and let $x_{ij}$ be the observed count in the $(i, j)$ cell and $m_{ij}$ be the corresponding expected value. For those cells not in the set $S$ we put $x_{ij} = m_{ij} = 0$ so that we can continue to use our usual notation for marginal totals. For example, if $S$ consists of all cells but (1, 1), then

$$m_{1+} = \sum_{j=1}^{J} m_{ij} = \sum_{j=2}^{J} m_{ij}$$

still represents the expected count for the first row of the incomplete array $S$.

We propose to use the same loglinear models for the incomplete array $S$ as we did in Chapter 2 for complete arrays; that is, by analogy with the analysis of variance, for cells $(i, j) \in S$ we set

$$\log m_{ij} = u + u_{1(i)} + u_{2(j)} + u_{12(ij)}, \tag{8.2}$$

where

$$\sum_{i=1}^{I} u_{1(i)} = \sum_{j=1}^{J} u_{2(j)} = 0 \tag{8.3}$$

and

$$\sum_{i=1}^{I} \delta_{ij} u_{12(ij)} = \sum_{j=1}^{J} \delta_{ij} u_{12(ij)} = 0, \tag{8.4}$$

with

$$\delta_{ij} = \begin{cases} 1 & \text{for } (i, j) \in S, \\ 0 & \text{otherwise.} \end{cases} \tag{8.5}$$

Those $u_{12}$-terms in (8.4) that correspond to cells not in $S$ are set equal to an arbitrary finite quantity so that (8.4) is well-defined, and they never enter into formal consideration.

A very special type of incomplete array arises when we observe objects or individuals in pairs, and the pair $(j, i)$ is indistinguishable from the pair $(i, j)$ (e.g., frequencies of amino acid allele pairs in a protein). The resulting data are usually displayed in a triangular table with the same labels for rows as for columns. For such tables it is natural to set $u_{1(i)} = u_{2(i)}$ in expression (8.2). Larntz and Weisberg [1976] give further details on the use of loglinear models for such situations.

We now define the model of *quasi-independence* by setting

$$u_{12(ij)} = 0 \quad \text{for} \quad (i, j) \in S, \tag{8.6}$$

so that

$$\log m_{ij} = u + u_{1(i)} + u_{2(j)} \quad \text{for} \quad (i, j) \in S. \tag{8.7}$$

In other words, the variables corresponding to rows and columns are quasi-independent if we can write the $\{m_{ij}\}$ in the form

$$m_{ij} = \begin{cases} a_i b_j & \text{for} \quad (i, j) \in S, \\ 0 & \text{otherwise.} \end{cases} \tag{8.8}$$

Quasi-independence is like independence as it applies to the nonempty cells of a table.

When we estimate the expected cell values in an incomplete table under the model of quasi-independence, the MLEs under all three of the usual sampling schemes are the same (see Chapters 2, 3, and 5 and 6) and are usually unaffected by the presence of sampling zeros, provided no row or column has an observed zero total (for technical details see Fienberg [1970c, 1972a], Haberman [1974a], and Savage [1973]). These MLEs are uniquely determined by setting observed marginal totals equal to the MLEs of these expectations:

$$\begin{aligned} \hat{m}_{i+} &= x_{i+}, \qquad i = 1, 2, \ldots, I, \\ \hat{m}_{+j} &= x_{+j}, \qquad j = 1, 2, \ldots, J, \end{aligned} \tag{8.9}$$

where the $\hat{m}_{ij}$ satisfy the constraints of the model—(8.7) or (8.8).

Only in some special cases (see Bishop, Fienberg, and Holland [1975]) can we solve the maximum likelihood equations in closed form; however, we can always get the solution by using the following version of iterative proportional fitting. For our initial estimates we take

$$\hat{m}_{ij}^{(0)} = \delta_{ij}, \tag{8.10}$$

where $\delta_{ij}$ is defined by (8.5). Then at the $\nu$th cycle of the iteration ($\nu > 0$) we take

$$\hat{m}_{ij}^{(2\nu+1)} = \hat{m}_{ij}^{(2\nu)} \frac{x_{i+}}{\hat{m}_{i+}^{(2\nu)}}, \tag{8.11}$$

$$\hat{m}_{ij}^{(2\nu+2)} = \hat{m}_{ij}^{(2\nu+1)} \frac{x_{+j}}{\hat{m}_{+j}^{(2\nu+1)}}, \tag{8.12}$$

continuing until the values for the $\{\hat{m}_{ij}\}$ converge. Note the similarity between (8.11) and (8.12) and (3.34)–(3.36), and their relationships to the maximum likelihood equations.

The general rule (3.42) for computing degrees of freedom is still applicable here. If there are $e$ cells that have missing or fixed entries, then the set $S$ contains $IJ - e$ cells. The number of parameters fitted are $I + J - 1$ (1 for $u$, $I - 1$ for the $u_{1(i)}$, $J - 1$ for the $u_{2(j)}$, leaving $IJ - e - (I + J - 1) = (I - 1)\cdot(J - 1) - e$ degrees of freedom. Alternatively, we can use expression (8.1), with $T_e = IJ$, $Z_e = e$, and $T_p = I + J - 1$.

In some situations it is desirable to break an incomplete set $S$ into parts that can considered separately for the calculation of d.f. Incomplete tables in which this is possible are referred to as *separable*, and the interested reader is referred to the papers cited above for the technical details. Separability is not a problem for the examples we consider here.

*Example:* Ploog [1967] observed the distribution of genital display among the members of a colony of six squirrel monkeys (labeled in Table 8-2a as $R$, $S$, $T$, $U$, $V$, and $W$). For each display there is an active and passive participant, but a monkey never displays toward himself. Before analyzing the data in Table 8-2a we delete the row for subject $T$, since its marginal total is zero, and thus all MLEs for cells in that row under quasi-independence are also zero.

Table 8-2b contains the MLEs under the model of quasi-independence, computed using the iterative procedure. The two goodness-of-fit statistics used to test the appropriateness of the model of quasi-independence are both extremely large: $X^2 = 168.05$ and $G^2 = 135.17$, each with 15 d.f. (there are 5 rows, 6 columns, and 5 a priori zero cells, yielding d.f. $= 4 \times 5 - 5 = 15$).

A simple examination of the observed and expected values reveals just how bad the fit is, and we conclude that various monkeys choose to display themselves more often toward specific members of the colony. This information is helpful in assessing the social structure in the colony.

**Table 8-2**
Distribution of Genital Display in a Colony of Six Squirrel Monkeys, as Reported by Ploog [1967]. Rows represent active participants and columns represent passive participants.

(a) Observed values

| Name | $R$ | $S$ | $T$ | $U$ | $V$ | $W$ | Total |
|---|---|---|---|---|---|---|---|
| $R$ | — | 1 | 5 | 8 | 9 | 0 | 23 |
| $S$ | 29 | — | 14 | 46 | 4 | 0 | 93 |
| $T$ | 0 | 0 | — | 0 | 0 | 0 | 0 |
| $U$ | 2 | 3 | 1 | — | 38 | 2 | 46 |
| $V$ | 0 | 0 | 0 | 0 | — | 1 | 1 |
| $W$ | 9 | 25 | 4 | 6 | 13 | — | 57 |
| Total | 40 | 29 | 24 | 60 | 64 | 3 | 220 |

(b) Expected values under model of quasi-independence

| Name | $R$ | $S$ | $T$ | $U$ | $V$ | $W$ | Total |
|---|---|---|---|---|---|---|---|
| $R$ | — | 5.26 | 2.48 | 8.22 | 6.65 | 0.40 | 23.01 |
| $S$ | 19.18 | — | 10.32 | 34.18 | 27.66 | 1.65 | 93.00 |
| $U$ | 10.94 | 12.47 | 5.88 | — | 15.77 | 0.94 | 46.00 |
| $V$ | 0.22 | 0.25 | 0.12 | 0.39 | — | 0.02 | 1.00 |
| $W$ | 9.66 | 11.02 | 5.20 | 17.21 | 13.92 | — | 57.01 |
| Total | 40.01 | 29.00 | 24.00 | 60.00 | 64.00 | 3.01 | 220.02 |

## 8.3 Incompleteness in Several Dimensions

When we deal with incomplete multidimensional tables, by analogy with the two-dimensional situation we can consider loglinear models applied only to cells whose values are not missing or determined a priori. For simplicity we look at three-dimensional tables since the extension to higher dimensions is straightforward. $S$ now represents the cells in an incomplete $I \times J \times K$ table with observed and expected values $x_{ijk}$ and $m_{ijk}$, respectively, and we set $x_{ijk} = m_{ijk} = 0$ for $(i, j, k) \notin S$ so that we can continue using our usual notation for marginal totals. The most general quasi-loglinear model for cells $(i, j, k)$ not in $S$ is

$$\log m_{ijk} = u + u_{1(i)} + u_{2(j)} + u_{3(k)} + u_{12(ij)} + u_{13(ik)} + u_{23(jk)} + u_{123(ijk)}, \tag{8.13}$$

where as usual the $u$-terms are deviations and sum to zero over each included variable. For example,

$$\sum_i \delta_i^{(23)} u_{1(i)} = \sum_i \delta_{ij}^{(3)} u_{12(ij)} = \sum_i \delta_{ik}^{(2)} u_{13(ik)}$$
$$= \sum_i \delta_{ijk} u_{123(ijk)} = 0, \tag{8.14}$$

with

$$\delta_{ijk} = \begin{cases} 1 & \text{if } (i, j, k) \in S, \\ 0 & \text{otherwise,} \end{cases}$$
$$\delta_{ij}^{(3)} = \begin{cases} 1 & \text{if } \delta_{ijk} = 1 \text{ for some } k, \\ 0 & \text{otherwise,} \end{cases} \tag{8.15}$$
$$\delta_i^{(23)} = \begin{cases} 1 & \text{if } \delta_{ijk} = 1 \text{ for some } (j, k), \\ 0 & \text{otherwise,} \end{cases}$$

and similar definitions for $\delta_{ik}^{(2)}$, $\delta_{jk}^{(1)}$, $\delta_i^{(13)}$, and $\delta_k^{(12)}$. We set those $u$-terms in (8.14) that are not included in (8.13) equal to an arbitrary finite quantity so that expression (8.13) is well-defined and we can continue using the same notation as in the analysis of complete multidimensional tables.

We define various unsaturated quasi-loglinear models by setting $u$-terms in (8.13) equal to zero, and, as in the analysis of complete tables, we restrict our attention to hierarchical models. Haberman [1974a] gives a detailed discussion regarding conditions (e.g., on the structure of $S$) that ensure the existence of unique nonzero MLEs (also see Bishop, Fienberg, and Holland [1975]).

Where there exist unique nonzero MLEs for the nonzero expected cells or an incomplete multidimensional table and a particular quasi-loglinear model, these MLEs are uniquely determined by setting expected marginal configurations equal to the observed marginal configurations corresponding to the minimal sufficient statistics. For example, suppose we are fitting the model specified by $u_{123} = 0$ in a three-dimensional table. Then the MLEs are given by

$$\hat{m}_{ij+} = x_{ij+}, \qquad \hat{m}_{i+k} = x_{i+k}, \qquad \hat{m}_{+jk} = x_{+jk}, \tag{8.16}$$

where the subscripts in each set of equations range over all sets of values for which the expected marginal values are positive.

We can use the same iterative proportional fitting algorithm to solve (8.16) as we did in the complete table case to solve (3.32). We take as our initial values

$$\hat{m}_{ijk}^{(0)} = \delta_{ijk}, \tag{8.17}$$

**Table 8-3**
Results From Survey of Teenagers Regarding Their Health Concerns (Brunswick [1971]): Cross-Classification by Sex, Age, and Health Concerns

| Health Concerns | Male 12–15 | Male 16–17 | Female 12–15 | Female 16–17 |
|---|---|---|---|---|
| Sex, Reproduction | 4 | 2 | 9 | 7 |
| Menstrual Problems | — | — | 4 | 8 |
| How Healthy I Am | 42 | 7 | 19 | 10 |
| Nothing | 57 | 20 | 71 | 31 |

where $\delta_{ijk}$ is defined in (8.15). Then we cycle through steps (3.34)–(3.36) just as in the case of complete tables.

The computation of degrees of freedom again follows from the general formula (8.1). Illustrations are given in the following examples.

*Example.* Brunswick [1971] reports the results of a survey inquiring into the health concerns of teenagers. Table 8-3 presents an excerpt of the data from this study previously analyzed by Grizzle and Williams [1972]. The three variables displayed are sex (male, female), age (broken into two categories), and health concerns (broken into four categories). Since males do not menstruate, there are two structural zeros in this table and a structural zero in the sex by health concern two-dimensional marginal table. Table 8-4 gives the log-likelihood ratios and Pearson chi-square values for various loglinear models fitted to these data. Note that the degrees of freedom are reduced by two from the usual degrees of freedom for a complete table, unless the sex vs. health concern margin is fitted, in which case the reduction is by one.

The fit of the models in which $u_{13} = 0$ is quite poor, while the fit of each of the models
(a) $u_{123} = 0$,
(b) $u_{12} = u_{123} = 0$,
(c) $u_{23} = u_{123} = 0$,
is acceptable at the 0.05 level of significance. Moreover, the difference in fit between models (a) and (b) is quite small, while the difference in fit between models (a) and (c) is almost significant at the 0.05 level of significance. This analysis would suggest that model (b) is an appropriate one for the data, its

**Table 8-4**
The Goodness-of-Fit for Various Loglinear Models as Applied to the Data in Table 8-3 (variable 1 = sex, variable 2 = age, variable 3 = health concerns)

| Model | $G^2$ | $X^2$ | d.f. |
|---|---|---|---|
| [12] [13] [23] | 2.03 | 2.03 | 2 |
| [13] [23] | 4.86 | 4.98 | 3 |
| [12] [23] | 13.45 | 13.12 | 4 |
| [12] [13] | 9.43 | 9.62 | 5 |
| [12] [3] | 22.03 | 22.59 | 7 |
| [13] [2] | 15.64 | 15.95 | 6 |
| [23] [1] | 17.46 | 17.05 | 5 |
| [1] [2] [3] | 28.24 | 30.53 | 8 |

interpretation being that, given a particular health concern (other than menstrual problems), there is no relationship between the age and sex of individuals with that concern. We note that this is not the model decided upon by Grizzle and Williams [1972]. They prefer model (a), but their test statistic for model (b) given model (a), which is quite different from ours in numerical value, appears to be in error.

*Example.* For our second example we return to the data in Table 3-6 on occupation, aptitude, and education. When we analyzed these data in Chapter 3, we chose to treat all of the zero entries as sampling zeros; however, the zeros in the first two columns of the subtable for occupation 3 (teacher) bear further consideration. Few teachers in public schools, high schools, and colleges have themselves not completed high school. The exceptions may well be "nonstandard" teachers (e.g., of karate or music), whose responses should perhaps not be analyzed with the remainder of the sample. We therefore reanalyze the data from Table 3-6 treating all the entries in these two columns as if they were a priori zero. One might also wish to treat the two zeros under occupation 2 (self-employed, professional) in a similar fashion; this task will be left as an exercise for the reader.

As before, we take variable 1 to be occupation, variable 2 aptitude, and variable 3 education. The fits of all eight loglinear models to the truncated data are given in Table 8-5, in a manner parallel to that of Table 3-7. Note the differences in the degrees of freedom.

The fit of the models differs little from our earlier analysis. The two models providing an acceptable fit to the data are:

(1) $u_{123} = 0$,

(2) $u_{12} = u_{123} = 0$.

**Table 8-5**
Values of Likelihood-Ratio Goodness-of-Fit Statistics for Data in Table 3-6 When Entries for E1 and E2 under O3 Are Treated as Empty A Priori

| Model | d.f. | $G^2$ |
|---|---|---|
| [12][13][23] | 28 | 17.3 |
| [12][13] | 40 | 183.7* |
| [13][23] | 40 | 41.8 |
| [12][23] | 35 | 867.3* |
| [12][3] | 47 | 1046.1* |
| [13][2] | 52 | 219.1* |
| [23][1] | 47 | 904.2* |
| [1][2][3] | 59 | 1081.6* |

*denotes values in upper 5% tail of corresponding $\chi^2$ distribution, with d.f. as indicated.

The conditional test of fit for model (2) given model (1) is based on $G^2 = 41.8 - 17.3 = 24.5$ with 12 d.f. This value, when referred to the $\chi^2_{12}$ distribution, corresponds to a descriptive level of significance somewhere between 0.05 and 0.01, as before. Again we prefer model (2) to model (1), in part because of its conditional-independence interpretation. Table 8-6 contains the fitted values for model (2). When we examine the standardized residuals of the form $(x_{ijk} - \hat{m}_{ijk})/\sqrt{\hat{m}_{ijk}}$, we find that the largest residuals correspond to the column E4 (the highest level of education), and contribute well over 50% of the value of $G^2$ for this model.

## 8.4 Some Applications of Incomplete Table Methodology

Not only are loglinear models for incomplete tables useful in their own right, they have also proved to be of special utility for a variety of seemingly unrelated problems such as (1) fitting separate models to different parts of a table, (2) examining square tables for various aspects of symmetry, (3) estimating the size of a closed population through the use of multiple censuses or labelings. These applications are described in great detail in Chapters 5, 8, and 6, respectively, of Bishop, Fienberg, and Holland [1975].

Several new applications of incomplete table methodology have recently appeared, and three of these are described below.

### The Bradley–Terry paired comparisons model

Suppose $t$ items (e.g., different types of chocolate pudding) or treatments, labeled $T_1, T_2, \ldots, T_t$, are compared in pairs by sets of judges. The Bradley–

**Table 8-6**
Estimated Expected Values under Quasi-Loglinear Model of Conditional Independence of Occupation and Aptitude Given Education

O1

| | E1 | E2 | E3 | E4 |
|---|---|---|---|---|
| A1 | 49.5 | 58.4 | 26.0 | 4.5 |
| A2 | 64.6 | 79.1 | 53.5 | 10.8 |
| A3 | 85.3 | 107.2 | 84.0 | 21.0 |
| A4 | 29.4 | 49.4 | 44.8 | 10.2 |
| A5 | 10.1 | 14.9 | 24.7 | 6.4 |

O2

| | E1 | E2 | E3 | E4 |
|---|---|---|---|---|
| A1 | 1.2 | 2.1 | 7.8 | 17.1 |
| A2 | 1.6 | 2.8 | 16.1 | 40.5 |
| A3 | 2.1 | 3.8 | 25.2 | 79.1 |
| A4 | 0.7 | 1.8 | 13.5 | 38.2 |
| A5 | 0.2 | 0.5 | 7.4 | 24.1 |

O3

| | E1 | E2 | E3 | E4 |
|---|---|---|---|---|
| A1 | — | — | 1.3 | 18.4 |
| A2 | — | — | 2.8 | 43.8 |
| A3 | — | — | 4.3 | 85.5 |
| A4 | — | — | 2.3 | 41.3 |
| A5 | — | — | 1.3 | 26.0 |

O4

| | E1 | E2 | E3 | E4 |
|---|---|---|---|---|
| A1 | 164.3 | 147.5 | 102.9 | 43.0 |
| A2 | 214.7 | 200.0 | 211.7 | 102.0 |
| A3 | 283.5 | 271.0 | 332.4 | 199.3 |
| A4 | 97.8 | 124.8 | 177.4 | 96.3 |
| A5 | 33.6 | 37.6 | 97.6 | 60.6 |

Terry model postulates that the probability of $T_i$ being preferred to $T_j$ is

$$\Pr(T_i \succ T_j) = \frac{\pi_i}{\pi_i + \pi_j}, \qquad \begin{array}{l} i \neq j, \\ i, j = 1, 2, \ldots, t, \end{array} \tag{8.18}$$

where each $\pi_i \geq 0$ and we add the constraint that $\Sigma_{i=1}^{t} \pi_i = 1$. The model assumes independence of the same pair by different judges and different pairs by the same judge.

In the typical paired comparison experiment, $T_i$ is compared with $T_j$ $n_{ij} \geq 0$ times, and we let $x_{ij}$ be the observed number of times $T_i$ is preferred to $T_j$ in these $n_{ij}$ comparisons. Table 8-7a shows the typical layout for the observed data when $t = 4$, with preference (for, against) defining rows and columns. An alternative layout is suggested by the binomial nature of the sampling scheme, that is, the fact that the $\{x_{ij}\}$ are observations on independent binomial random variables with sample sizes $\{n_{ij}\}$ and "success" probabilities $\{\pi_i/(\pi_i + \pi_j)\}$. In this second layout we cross-classify the counts by the preferred treatment and by the pair being compared. Thus we get an incomplete $t \times \binom{t}{2}$ table, illustrated for the case $t = 4$ in Table 8-7b.

The column totals in Table 8-7b are now fixed by design, and the Bradley–Terry model of expression (8.18) is identical to the model of quasi-independence (see expression (8.8)) for the corresponding table of expected values once we take these fixed totals into account. Fienberg and Larntz [1976]

**Table 8-7**
Two Layouts for Data in Paired-Comparisons Study with $t = 4$

(a) Typical layout

| | | Against | | | |
|---|---|---|---|---|---|
| | | $T_1$ | $T_2$ | $T_3$ | $T_4$ |
| For | $T_1$ | — | $x_{12}$ | $x_{13}$ | $x_{14}$ |
| | $T_2$ | $x_{21}$ | — | $x_{23}$ | $x_{24}$ |
| | $T_3$ | $x_{31}$ | $x_{32}$ | — | $x_{34}$ |
| | $T_4$ | $x_{41}$ | $x_{42}$ | $x_{43}$ | — |

(b) Alternative layout

| | | Paired Comparison No. | | | | | |
|---|---|---|---|---|---|---|---|
| | | 1 | 2 | 3 | 4 | 5 | 6 |
| Preferred Treatment | $T_1$ | $x_{12}$ | $x_{13}$ | $x_{14}$ | — | — | — |
| | $T_2$ | $x_{21}$ | — | — | $x_{23}$ | $x_{24}$ | — |
| | $T_3$ | — | $x_{31}$ | — | $x_{32}$ | — | $x_{34}$ |
| | $T_4$ | — | — | $x_{41}$ | — | $x_{42}$ | $x_{43}$ |

give further details regarding loglinear representations for the Bradley–Terry model and its generalizations.

One of the earliest applications of paired comparison models was to the analysis of win–loss data for major league baseball teams (see Mosteller [1951]). We illustrate the use of the Bradley–Terry model here on the American League baseball records for 1975. The observed data are given in Table 8-8, in a form analogous to Table 8-7a in order to conserve space. The American League consists of two divisions, and teams are scheduled to play 18 games with the other 5 teams in the same division, and 12 games with the remaining 6 teams. The total number of games played may be less than 12 or 18 because of cancellation of games at the end of the season. In one case (Milwaukee–New York) the total is 19—clearly an error, but one we choose to overlook.

The Bradley–Terry model gives a very good fit to the data in Table 8-8: $G^2 = 56.6$ with 55 d.f. The estimated expected values under this model are given in Table 8-9. An examination of the standaridized residuals, $(x_{ij} - \hat{m}_{ij})/\hat{m}_{ij}$, shows no aberrant cells.

When Mosteller [1951] applied a similar model to 1948 baseball data, he also found a remarkably good fit. One possible explanation for the good fits is that data like those in Table 8-8 are a mixture of games played at home and away. Were we able to separate out these two layers, we might find that

**Table 8-8**
Wins and Losses for American League Baseball Clubs, 1975 (unofficial) (read across table for wins and down for losses)

| | Balt. | Bos. | Cal. | Chi. | Cle. | Det. | Kan. | Mil. | Minn. | N. Y. | Oak. | Tex. |
|---|---|---|---|---|---|---|---|---|---|---|---|---|
| Balt. | — | 9 | 6 | 7 | 10 | 12 | 7 | 13 | 6 | 8 | 4 | 7 |
| Bos. | 9 | — | 6 | 8 | 5 | 12 | 7 | 10 | 10 | 11 | 6 | 8 |
| Cal. | 6 | 6 | — | 7 | 3 | 6 | 4 | 7 | 7 | 7 | 7 | 9 |
| Chi. | 4 | 4 | 9 | — | 7 | 5 | 9 | 8 | 9 | 6 | 9 | 5 |
| Cle. | 8 | 11 | 9 | 5 | — | 12 | 6 | 9 | 3 | 8 | 2 | 5 |
| Det. | 4 | 5 | 5 | 7 | 5 | — | 6 | 7 | 4 | 6 | 6 | 1 |
| Kan. | 5 | 5 | 14 | 9 | 6 | 6 | — | 7 | 11 | 7 | 6 | 13 |
| Mil. | 4 | 8 | 5 | 4 | 9 | 10 | 5 | — | 2 | 9 | 5 | 6 |
| Minn. | 6 | 2 | 10 | 7 | 6 | 11 | 7 | 9 | — | 4 | 6 | 8 |
| N. Y. | 10 | 5 | 5 | 6 | 7 | 12 | 4 | 10 | 8 | — | 6 | 8 |
| Oak. | 8 | 6 | 10 | 9 | 10 | 6 | 10 | 9 | 12 | 6 | — | 9 |
| Tex. | 5 | 4 | 9 | 12 | 7 | 11 | 4 | 6 | 10 | 4 | 5 | — |

the Bradley–Terry parameters associated with winning at home and winning away are different. Then collapsing the data to form Table 8-8 would tend to constrain the variability in the resulting counts, yielding a better fit than we could expect even if the Bradley–Terry model were appropriate.

**Partitioning polytomous variables: Collapsing**

Often, when analyzing cross-classifications involving one or more polytomous variables, we wish to collapse these variables as much as possible for ease of interpretation, or we wish to infer how specific categories of a polytomy interact with other variables. In Chapter 3 we gave a rule for when it is possible to collapse across variables, and this same rule is applicable for collapsing across categories within a variable (see Bishop, Fienberg, and Holland [1975], Chapter 2). A formal method for checking on collapsibility suggested by Duncan [1975b] involves representing a complete cross-classification in the form of an incomplete one by replacing the polytomous variable with a series of dichotomous variables. We illustrate this method with an example taken from Duncan [1975b].

Table 8-10 contains data in the form of a three-dimensional cross-classi-

**Table 8-9**
Expected Wins and Losses for Data in Table 8-8 under Bradley-Terry Paired Comparisons Model (read across table for wins and down for losses)

| | Balt. | Bos. | Cal. | Chi. | Cle. | Det. | Kan. | Mil. | Minn. | N. Y. | Oak. | Tex. |
|---|---|---|---|---|---|---|---|---|---|---|---|---|
| Balt. | — | 8.6 | 7.3 | 6.3 | 9.9 | 11.0 | 6.0 | 10.8 | 6.9 | 9.9 | 5.5 | 6.8 |
| Bos. | 9.4 | — | 7.5 | 7.1 | 9.1 | 12.0 | 6.2 | 11.8 | 7.1 | 9.1 | 5.8 | 7.0 |
| Cal. | 4.7 | 4.5 | — | 7.4 | 5.3 | 6.5 | 7.0 | 6.3 | 7.9 | 5.2 | 6.0 | 8.2 |
| Chi. | 4.7 | 4.9 | 8.6 | — | 5.7 | 7.5 | 7.7 | 6.8 | 8.0 | 5.7 | 7.0 | 8.4 |
| Cle. | 8.1 | 6.9 | 6.7 | 6.3 | — | 10.9 | 5.4 | 10.6 | 4.7 | 7.4 | 4.9 | 6.2 |
| Det. | 5.0 | 5.0 | 4.5 | 4.5 | 6.1 | — | 3.7 | 7.5 | 5.7 | 6.3 | 3.3 | 4.4 |
| Kan. | 6.0 | 5.8 | 11.0 | 10.3 | 6.6 | 8.3 | — | 7.6 | 10.4 | 6.0 | 7.4 | 9.6 |
| Mil. | 6.2 | 6.2 | 5.7 | 5.2 | 7.4 | 9.5 | 4.4 | — | 4.8 | 7.8 | 4.6 | 5.1 |
| Minn. | 5.1 | 4.9 | 9.1 | 8.0 | 4.3 | 9.3 | 7.6 | 6.2 | — | 5.7 | 7.0 | 8.8 |
| N. Y. | 8.1 | 6.9 | 6.8 | 6.3 | 7.6 | 11.7 | 5.0 | 11.2 | 6.3 | — | 5.0 | 6.2 |
| Oak. | 6.5 | 6.2 | 11.0 | 11.0 | 7.1 | 8.7 | 8.6 | 9.4 | 11.0 | 7.0 | — | 8.4 |
| Tex. | 5.2 | 5.0 | 9.8 | 8.6 | 5.8 | 7.6 | 7.4 | 6.9 | 9.2 | 5.8 | 5.6 | — |

fication of responses to a question, addressed to mothers of children under 19 years of age, concerning whether boys, girls, or both should be required to shovel snow from sidewalks. Since no mother gave a "girl" response, we have listed the data using a dichotomy, labeled here as variable 1 (boy, both). The two explanatory variables in this study are year (1953, 1971) and religion (Protestant, Catholic, Jewish, Other), labeled as variables 2 and 3, respectively.

Table 8-11 lists the four loglinear models that include the $u_{23}$-term corresponding to the explanatory variables and the corresponding likelihood-ratio goodness-of-fit statistics and degrees of freedom. Clearly the no second-order interaction model fits the data (perhaps too well!). Although the model that includes an effect of year on response ([12] [23]) also fits the data moderately well, the conditional test for this model given the no second-order interaction model, $G^2 = 11.2 - 0.4 = 10.8$ with 3 d.f., has a descriptive level of significance somewhere between 0.05 and 0.01.

Perhaps the effect of religion on the response, which seems necessary for the model, can be accounted for by a single religious category. If this is the case, we can collapse the religion variable and get a more parsimonious

**Table 8-10**
Observed Frequencies of Response to "Shoveling" Question, by Religion by Year (Duncan [1975b])

| | Year | | | |
|---|---|---|---|---|
| | 1953 | | 1971 | |
| Religion | Boy | Both | Boy | Both |
| Protestant | 104 | 42 | 165 | 142 |
| Catholic | 65 | 44 | 100 | 130 |
| Jewish | 4 | 3 | 5 | 6 |
| Other | 13 | 6 | 32 | 23 |

**Table 8-11**
Loglinear Models Fit to Data in Table 8-10, and Their Likelihood-Ratio Goodness-of-Fit statistics (All Models Include $u_{23}$)

| Model | $G^2$ | d.f. |
|---|---|---|
| 1. [1] [23] | 31.7 | 7 |
| 2. [12] [23] | 11.2 | 6 |
| 3. [13] [23] | 21.5 | 4 |
| 4. [12] [13] [23] | 0.4 | 3 |

and compact summary of the data. To explore this issue we replace variable 3 (religion) by a series of four dichotomous variables, labeled 4 through 7, corresponding to Protestant, Catholic, Jewish, or Other, respectively. Each variable takes the value 1 if the respondent has that religious affiliation and 0 otherwise. The redundancy introduced by using four rather than three dichotomies allows us to treat the four categories symmetrically. Now we can display the data in the form of a six-dimensional incomplete array, as indicated in Table 8-12. Models corresponding to those fit above must keep $u_{24567}$ fixed, and in Table 8-13 we include four such models that are intermediate to the two models with acceptable fits. No formal analysis is required to show that the interaction between religion and response is due primarily to Catholic respondents. The fitted values for this model are given in Table 8-12. Thus we can collapse religion in this example to a dichotomy.

### Scaling response patterns

In psychometric and sociometric investigations, subjects are often classified according to their responses on a sequence of $p$ dichotomous items (all of which have categories yes and no). These items are said to form a *perfect Guttman scale* if we can reproduce exactly an individual's response from the

**Table 8-12**
Estimated Expected Values for Model 7 in Table 8-13, with Effects of Year and Religion-Catholic on the Response Variable (Duncan [1975b])

| Formal Variables for Religion | | | | Year | | | |
|---|---|---|---|---|---|---|---|
| | | | | 1953 | | 1971 | |
| Protestant | Catholic | Jewish | Other | Boy | Both | Boy | Both |
| 1 | 1 | 1 | 1 | — | — | — | — |
| 1 | 1 | 1 | 0 | — | — | — | — |
| 1 | 1 | 0 | 1 | — | — | — | — |
| 1 | 1 | 0 | 0 | — | — | — | — |
| 1 | 0 | 1 | 1 | — | — | — | — |
| 1 | 0 | 1 | 0 | — | — | — | — |
| 1 | 0 | 0 | 1 | — | — | — | — |
| 1 | 0 | 0 | 0 | 102.31 | 43.69 | 166.65 | 140.35 |
| 0 | 1 | 1 | 1 | — | — | — | — |
| 0 | 1 | 1 | 0 | — | — | — | — |
| 0 | 1 | 0 | 1 | — | — | — | — |
| 0 | 1 | 0 | 0 | 65.47 | 43.53 | 99.53 | 130.47 |
| 0 | 0 | 1 | 1 | — | — | — | — |
| 0 | 0 | 1 | 0 | 4.91 | 2.09 | 5.97 | 5.03 |
| 0 | 0 | 0 | 1 | 13.31 | 5.69 | 29.86 | 25.14 |
| 0 | 0 | 0 | 0 | — | — | — | — |

**Table 8-13**
Loglinear Models Fit to Six-Dimensional Version of Data in Table 8-10 (All Models Include $u_{24567}$)

| Model | $G^2$ | d.f. |
|---|---|---|
| 2.* [24567] [12] | 11.2 | 6 |
| 4.* [24567] [12] [14567] | 0.4 | 3 |
| 5. [24567] [12] [17] | 9.8 | 5 |
| 6. [24567] [12] [16] | 10.9 | 5 |
| 7. [24567] [12] [15] | 1.4 | 5 |
| 8. [24567] [12] [14] | 4.8 | 5 |

* Models 2 and 4 from Table 8-11.

number of "yes" responses to the $P$ items. This is equivalent to saying that the items have an order and that there are exactly $p + 1$ possible responses for individuals. If $p = 4$ and $1 =$ yes while $2 =$ no, the five response patterns are

(1) (1, 1, 1, 1)
(2) (1, 1, 1, 2)

(3) (1, 1, 2, 2)
(4) (1, 2, 2, 2)
(5) (2, 2, 2, 2).

In practice, when we examine samples of individuals, we rarely find that the items form a perfect Guttman scale. We refer to individuals with response patterns other than those listed as being "unscalable." Goodman [1975] has proposed the following model for dealing with data in which the items fall short of forming a perfect Guttman scale:

(i) All individuals can be classified into one of $p + 2$ categories. The 0th category is for "unscalable" individuals, and categories 1 through $p + 1$ correspond to the $p + 1$ Guttman scale types.
(ii) The responses of the unscalable individuals on the $p$ items are completely independent.
(iii) The responses of the scalable individuals correspond to their scale type.

Note that point (ii) allows unscalable individuals to be observed as giving a scalable response by chance. For $p = 4$, if $p_{ijkl}$ is the probability of response $(i, j, k, l)$, this model can be written algebraically as

$$\begin{aligned} p_{1111} &= \pi_1 + \pi_0 a_1 b_1 c_1 d_1, \\ p_{1112} &= \pi_2 + \pi_0 a_1 b_1 c_1 d_2, \\ p_{1122} &= \pi_3 + \pi_0 a_1 b_1 c_2 d_2, \\ p_{1222} &= \pi_4 + \pi_0 a_1 b_2 c_2 d_2, \\ p_{2222} &= \pi_5 + \pi_0 a_2 b_2 c_2 d_2, \end{aligned} \tag{8.19}$$

and for all other response patterns

$$p_{ijkl} = \pi_0 a_i b_j c_k d_l, \tag{8.20}$$

where $\pi_i$ is the probability of an individual belonging to scale category $i$ and

$$\sum_{i=0}^{5} \pi_i = 1, \qquad a_1 + a_2 = b_1 + b_2 = c_1 + c_2 = d_1 + d_2 = 1. \tag{8.21}$$

Let us denote by $S$ the patterns corresponding to only unscalable individuals. If we look only at cells in $S$, then (8.20) implies that the four scale items are quasi-independent. To compute MLEs of the expected cell counts we fit the model of quasi-independence to the cells in $S$, replacing the other five cells with structural zeros, yielding

$$\hat{m}_{ijkl} = \begin{cases} N(\hat{\pi}_0 \hat{a}_i \hat{b}_j \hat{c}_k \hat{d}_l) & \text{for} \quad (i, j, k, l) \in S, \\ x_{ijkl} & \text{for} \quad (i, j, k, l) \notin S. \end{cases} \tag{8.22}$$

We can compute the estimates of $\pi_0$, $\{a_i\}$, $\{b_j\}$, $\{c_k\}$, and $\{d_l\}$ directly from (8.22) by looking at various ratios of estimated expected values. Then we use (8.21) to compute $\hat{\pi}_i$ for $i = 1, 2, 3, 4, 5$; for example,

$$\hat{\pi}_1 = N^{-1}(\hat{m}_{1111} - \hat{\pi}_0 \hat{a}_1 \hat{b}_1 \hat{c}_1 \hat{d}_1). \tag{8.23}$$

In the event that $\hat{\pi}_i < 0$ for some scalable individuals, we redo the calculation setting these estimated probabilities equal to zero, and we include the corresponding cells in the set $S$, fitted by the model of quasi-independence.

If there are $p$ dichotomous scale items and $\hat{\pi}_i > 0$ for all $i$, then the degrees of freedom for testing the goodness-of-fit of Goodman's model are

$$2^p - 2(p + 1). \tag{8.24}$$

We illustrate the application of this model for scaling response patterns on a data set used by Goodman [1975] and attributed by him to Stouffer and Toby [1951]. The data consist of responses of 216 individuals according to whether they tend toward *universalistic* values or *particularistic* values when

**Table 8-14**
Observed and Estimated Counts for Scaling Model Fit to Stouffer–Toby Data on Role Conflict (Goodman [1975])

| Response Pattern | | | | | Estimated |
|---|---|---|---|---|---|
| Items | | | | Observed | Expected |
| 1 | 2 | 3 | 4 | Counts | Counts |
| 1 | 1 | 1 | 1 | 42 | 42 |
| 1 | 1 | 1 | 2 | 23 | 23 |
| 1 | 1 | 2 | 1 | 6 | 4.72 |
| 1 | 1 | 2 | 2 | 25 | 25 |
| 1 | 2 | 1 | 1 | 6 | 5.99 |
| 1 | 2 | 1 | 2 | 24 | 24.74 |
| 1 | 2 | 2 | 1 | 7 | 7.56 |
| 1 | 2 | 2 | 2 | 38 | 38 |
| 2 | 1 | 1 | 1 | 1 | 1.14 |
| 2 | 1 | 1 | 2 | 4 | 4.73 |
| 2 | 1 | 2 | 1 | 1 | 1.44 |
| 2 | 1 | 2 | 2 | 6 | 5.97 |
| 2 | 2 | 1 | 1 | 2 | 1.83 |
| 2 | 2 | 1 | 2 | 9 | 7.57 |
| 2 | 2 | 2 | 1 | 2 | 2.31 |
| 2 | 2 | 2 | 2 | 20 | 20 |

confronted by each of four situations of role conflict (items 1–4). The raw counts and the estimated expected values are given in Table 8-14. The estimated expected values for the five scale types are equal to those observed. The fit of the model is extremely good: $G^2 = 0.99$ with 6 d.f. The estimated parameters in the model are

$$(\hat{\pi}_0, \hat{\pi}_1, \hat{\pi}_2, \hat{\pi}_3, \hat{\pi}_4, \hat{\pi}_5) = (0.69, 0.18, 0.03, 0.03, 0.03, 0.05)$$

and

$$\hat{a}_1 = 0.77, \quad \hat{b}_1 = 0.38, \quad \hat{c}_1 = 0.51, \quad \hat{d}_1 = 0.31.$$

Goodman [1975] not only describes the basic scaling model outlined here but also describes various extensions and relates the scaling model to a latent class model.

# Appendix I
# Statistical Terminology

Although this book is written primarily for nonstatisticians, occasionally statistical words are used that will be unfamiliar to readers whose introduction to statistics has come via a noncalculus methodology course or its equivalent. For those readers the meaning of terms such as *asymptotic*, *maximum-likelihood estimation*, *minimal sufficient statistics*, and *descriptive level of significance* may be unclear. While this appendix should not be viewed as a primer on theoretical statistical concepts, it may help some readers to move over what might otherwise be viewed as statistical stumbling blocks in the remainder of the monograph. For complete details the reader is referred to books such as Cox and Hinkley [1974] or Rao [1973].

The analysis of multidimensional contingency tables via loglinear models aims to make inferences about a set of parameters describing the structural relationship among the underlying variables. Suppose the observations $\boldsymbol{y} = (y_1, \ldots, y_n)$ are the realized values of a corresponding set of random variables $\boldsymbol{Y} = (Y_1, \ldots, Y_n)$. The *probability density function*, $f(\boldsymbol{y})$, of $\boldsymbol{Y}$ gives the frequency of observed $\boldsymbol{y}$ values. It is a function of $\boldsymbol{y}$ given the unknown parameters $\boldsymbol{\theta} = (\theta_1, \ldots, \theta_t)$; that is, $f(\boldsymbol{y}) = f(\boldsymbol{y}|\boldsymbol{\theta})$. The *likelihood function* of the observed data views $f(\boldsymbol{y}|\boldsymbol{\theta})$ as a function of $\boldsymbol{\theta}$ given a particular value of $\boldsymbol{y}$. The method of *maximum likelihood* finds the value of $\boldsymbol{\theta}$ that maximizes this likelihood function, and this value is labeled $\hat{\boldsymbol{\theta}}$. The most convenient way to find the maximum is to solve the *likelihood equations* found by taking the first derivatives of the likelihood function with respect to the parameters $\theta_1, \ldots, \theta_t$:

$$\frac{\partial f(\boldsymbol{y}|\boldsymbol{\theta})}{\partial \theta_i} = 0, \qquad i = 1, 2, \ldots t.$$

For the models used throughout this monograph the solution to the likelihood equations, when it exists, is unique.

Maximum-likelihood estimates have a variety of desirable properties, most of which are related to *asymptotic theory*, which corresponds to the situation where the sample size $N$ goes to infinity. Since asymptotic theory is associated with large samples, we often speak of large-sample distributions when we use asymptotic results. Three key asymptotic properties of maximum-likelihood estimates (MLEs) are of interest here:

(1) MLEs are *consistent:* if the entire population of random variables is observed, the MLEs are exactly equal to the parameters.

(2) MLEs are *asymptotically normally* distributed, with the true parameter values as the means and a covariance matrix that is relatively easy to calculate.

(3) MLEs are *asymptotically efficient:* it is impossible to produce other estimates with smaller asymptotic variances.

For these properties to hold, the probability function $f(\mathbf{y}|\theta)$ must satisfy certain smoothness properties. This is the case for all probability functions used in this monograph.

The asymptotic normality of property (2) leads to the asymptotic $\chi^2$ distribution of the test statistics used to check on the goodness-of-fit of loglinear models to a set of observed counts (see Bishop, Fienberg, and Holland [1975], Chapter 14).

Suppose the random variable $T$ is a test statistic used for the test of the null hypothesis $H_0$. Then the *descriptive level of significance* associated with an observed value of $T$, say $t$ (e.g., the observed value of the Pearson chi-square statistic), is

$$\Pr(T \geq t \mid H_0 \text{ true}),$$

and as such we use it as a measure of the consistency of the data with $H_0$.

# Appendix II
# Basic Estimation Results for Loglinear Models

Most of the methods used in this book are based on four results dealing with maximum-likelihood estimation of parameters in loglinear models. While the proofs of these results are somewhat complex (see, for example, Haberman [1974a]) and often rely on more general results about the existence and uniqueness of MLEs in exponential family distributions (see, for example, Andersen [1974]), their statements are quite simple. This appendix presents a brief summary of this general loglinear-model theory.

Since we often wish to consider ensembles of cells in many dimensions, and occasionally of irregular shape, we shall use single subscripts to describe counts. All of the problems are concerned with a set of *observed counts*,

$$\boldsymbol{x} = \{x_i : i \in \mathscr{I}\},$$

indexed by a set $\mathscr{I}$ containing $t$ elements, which are realizations of a set of random variables $\boldsymbol{X}$, similarly indexed. Corresponding to the observed counts are sets of expected values

$$\boldsymbol{m} = \{m_i = E(X_i) : i \in \mathscr{I}\}$$

and of log expected values

$$\boldsymbol{\lambda} = \{\lambda_i = \log m_i : i \in \mathscr{I}\}.$$

There are two sampling schemes used to describe the observed counts, $\boldsymbol{x}$. The counts are either (a) observations from independent Poisson distributions, or (b) observations from independent multinomial distributions. When sampling scheme (a) is used, the likelihood function is proportional to

$$\prod_{i \in \mathscr{I}} m_i^{x_i} e^{-m_i},$$

and thus the kernel of the log likelihood is

$$\sum_i x_i \log m_i - \sum_i m_i = \sum_i x_i \lambda_i - \sum_i m_i.$$

We are interested in situations where $\boldsymbol{m}$ is described by a loglinear model—that is, a linear model for $\boldsymbol{\lambda}$—which we denote by $M$.

**Result 1. Under Poisson sampling, corresponding to each parameter in $M$ is a minimal sufficient statistic that is expressible as a linear combination of the $\{x_i\}$.**

We denote the minimal sufficient statistics by $P_M\boldsymbol{x}$ ($P_M$ is the projection onto the linear subspace of the $t$-dimensional space of real numbers corresponding to the loglinear model $M$).

*Example.* Suppose the $\boldsymbol{x}$ form an $I \times J$ contingency table, $\{x_{ij}\}$, and $M$ corresponds to the model specifying independence of rows and columns. Then

$$P_M\boldsymbol{x} = \{x_{i+} : i = 1, 2, \ldots, I;\ x_{+j} : j = 1, 2, \ldots, J\}.$$

**Result 2. The MLE, $\hat{\boldsymbol{m}}$, of $\boldsymbol{m}$ under the loglinear model $M$ exists and is unique if and only if the likelihood equations,**

$$P_M\hat{\boldsymbol{m}} = P_M\boldsymbol{x},$$

**have a solution that is in the interior of the subspace corresponding to $M$.**

Note that the form of the likelihood equations is extremely simple. We take the minimal sufficient statistics, which from Result 1 are linear combinations of the $\{x_i\}$, and set them equal to their expectations. Determining when the solution of the equations lies within the interior of the subspace can be a tricky matter. A sufficient (but not necessary) condition for a proper solution to exist is $\boldsymbol{x} > 0$. For a further discussion of this matter see Haberman [1974a].

*Example* (continued). For the $I \times J$ table, the likelihood equations are

$$\begin{aligned} \hat{m}_{i+} &= x_{i+}, \qquad i = 1, 2, \ldots, I, \\ \hat{m}_{+j} &= x_{+j}, \qquad j = 1, 2, \ldots, J, \end{aligned}$$

which have the unique solution

$$m_{ij} = \frac{x_{i+}x_{+j}}{x_{++}}, \qquad \begin{array}{l} i = 1, 2, \ldots, I, \\ j = 1, 2, \ldots, J. \end{array}$$

Now suppose we break the $t$ cells in $\mathscr{I}$ up into $r$ sets. If the

$$\{x_i : i \in \mathscr{I}_k\}, \qquad k = 1, 2, \ldots, r,$$

now represent observations from $r$ independent multinomial distributions with

$$n_k = \sum_{i \in \mathscr{I}_k} x_i, \qquad k = 1, 2, \ldots, r,$$

fixed by design, we have a *product-multinomial* sampling, with log likelihood

$$\sum_i x_i \log m_i = \sum_i x_i \lambda_i,$$

subject to the constraints

$$n_k = \sum_{i \in \mathscr{I}_k} m_i, \qquad k = 1, 2, \ldots, r.$$

Note that the product-multinomial log likelihood is the first term in the Poisson log likelihood. When $r = 1$, we have the special case of *multinomial sampling*, where only the total sample size $N = \Sigma_{i \in \mathscr{I}} x_i$ is fixed.

Suppose $M$ is a loglinear model for a Poisson sampling scheme that includes among its sufficient statistics

$$n_k = \sum_{i \in \mathscr{I}_k} x_i, \qquad k = 1, 2, \ldots, r.$$

If we let $L$ correspond to a linear subspace defined by these $k$ constraints, and if we denote by $M \ominus L$ the part of the model $M$ that is appropriate for data generated by product-multinomial sampling, then we have:

**Result 3. The MLE of $\boldsymbol{m}$ under this product-multinomial sampling scheme for the model $M \ominus L$ is the same as the MLE of $\boldsymbol{m}$ under Poisson sampling for the model $M$.**

This key result on the equivalence of MLEs under different sampling schemes is used many times in the book.

*Example* (continued). In the $I \times J$ table, if $r = 1$, we have a single multinomial for the $IJ$ cells; if $r = I$, we have $I$ independent multinomials, one for each row; if $r = J$, we have $J$ independent multinomials, one for each column.

In all these cases, the MLEs under $M \ominus L$, where $M$ is the independence model, are

$$\hat{m}_{ij} = \frac{x_{i+}x_{+j}}{x_{++}}, \qquad \begin{array}{l} i = 1, 2, \ldots, I, \\ j = 1, 2, \ldots, J. \end{array}$$

We have chosen to describe the first three results in a coordinate-free fashion, although we also can describe them in terms of a specific coordinate system, a specific parametrization for $M$ and $M \ominus L$, and matrix notation. For example, in the case of a single multinomial we can write the model $M \ominus L$ as

$$\boldsymbol{m} = \frac{\{\boldsymbol{exp}(\boldsymbol{A\beta})\}}{\{\boldsymbol{1}_t'[\boldsymbol{exp}(\boldsymbol{A\beta})]\}},$$

where $\boldsymbol{A}$ is an appropriate $t \times s$ design matrix of known coefficients whose $s$ columns are linearly independent (of each other and of $\boldsymbol{1}_t$) and represent a basis for the subscripted parameters in the loglinear model, $\boldsymbol{\beta}$ is the corresponding $s \times 1$ vector of unknown parameters, $\boldsymbol{1}_t$ is a $t \times 1$ vector of 1s, and $\boldsymbol{exp}$ transforms a vector elementwise to the corresponding vector of exponential functions (see Koch, Freeman, and Tolley [1975]). The likelihood equations can then be expressed as

$$\boldsymbol{A}'\hat{\boldsymbol{m}} = \boldsymbol{A}'\boldsymbol{x}.$$

Using standard large-sample theory, we can estimate the asymptotic covariance matrix of either $\hat{\boldsymbol{\beta}}$ or $\hat{\boldsymbol{m}}$:

**Result 4. Under the multinomial sampling scheme an estimate of the asymptotic covariance matrix of $\hat{\boldsymbol{\beta}}$ is given by**

$$\{\boldsymbol{A}'[\boldsymbol{D}_{\hat{\boldsymbol{m}}} - \hat{\boldsymbol{m}}\hat{\boldsymbol{m}}'/N]\boldsymbol{A}\}^{-1},$$

**where $\boldsymbol{D}_{\hat{\boldsymbol{m}}}$ is a diagonal matrix with the elements of $\hat{\boldsymbol{m}}$ on the main diagonal. The corresponding estimated asymptotic covariance matrix of $\hat{\boldsymbol{m}}$ (under the model $M \ominus L$) is**

$$\{[\boldsymbol{D}_{\hat{\boldsymbol{m}}} - \hat{\boldsymbol{m}}\hat{\boldsymbol{m}}'/N]\boldsymbol{A}\}\,\{\boldsymbol{A}'[\boldsymbol{D}_{\hat{\boldsymbol{m}}} - \hat{\boldsymbol{m}}\hat{\boldsymbol{m}}'/N]\boldsymbol{A}\}^{-1}\,\{\boldsymbol{A}'[\boldsymbol{D}_{\hat{\boldsymbol{m}}} - \hat{\boldsymbol{m}}\hat{\boldsymbol{m}}'/N]\}.$$

Koch, Freeman, and Tolley [1975] also give formulas for asymptotic covariances under product-multinomial sampling. For both sampling schemes their results are equivalent to the closed-form expressions derived by Lee [1975] for loglinear models with direct, closed-form estimated expected values.

# Appendix III
# Percentage Points of $\chi^2$ Distribution

Detailed tables giving percentage points or tail errors of the $\chi^2$ distribution are readily available (e.g., Pearson and Hartley [1966]). Rather than present such a table here we give an abbreviated table plus a simple formula for computing the percentage points in the upper tail of the $\chi^2$ distribution, due to Hoaglin [1977]. If we let $\chi^2_{v,\alpha}$ be the value of the chi-square distribution with $v$ d.f. having an upper-tail probability of $\alpha$, then Hoaglin shows that

$$\sqrt{\chi^2_{v,\alpha}} \doteq \sqrt{v} + 2\sqrt{-\log_{10}\alpha} - \tfrac{7}{6}.$$

This formula has a percentage error for $\alpha = 0.05, 0.025, 0.01, 0.005$ of less than 1% for $6 < v \leq 100$. For $\alpha = 0.1$ the percentage error is less than 1.1%, and for $\alpha = 0.001$ it is less than 1.3%.

Line 7 of Table A-1 makes use of the Hoaglin formula.

When $v$ is large, the chi-square distribution percentage points can be well approximated by

$$\chi^2_{v,\alpha} = \tfrac{1}{2}(\sqrt{2v-1} + z_\alpha)^2,$$

where $z_\alpha$ is the value of the normal distribution, with mean 0 and variance 1, having an upper-tail probability of $\alpha$.

**Table A-1**
$\chi^2_{v,\alpha}$

| | | $\alpha$ | | | | |
|---|---|---|---|---|---|---|
| | | 0.1 | 0.05 | 0.025 | 0.01 | 0.005 |
| | 1 | 2.71 | 3.84 | 5.02 | 6.63 | 7.88 |
| | 2 | 4.61 | 5.99 | 7.38 | 9.21 | 10.60 |
| | 3 | 6.25 | 7.81 | 9.35 | 11.34 | 12.84 |
| $v$ | 4 | 7.78 | 9.49 | 11.14 | 13.28 | 14.86 |
| | 5 | 9.24 | 11.07 | 12.83 | 15.09 | 16.75 |
| | 6 | 10.64 | 12.59 | 14.45 | 16.81 | 18.55 |
| $v > 6$ | | $(\sqrt{v} + 0.833)^2$ | $(\sqrt{v} + 1.115)^2$ | $(\sqrt{v} + 1.365)^2$ | $(\sqrt{v} + 1.662)^2$ | $(\sqrt{v} + 1.867)^2$ |

# References


Andersen, A. H. [1974]. Multi-dimensional contingency tables. *Scand. J. Statist.* **1**, 115–127.

Anderson, J. B. and Davidovits, P. [1975]. Isotope separation in a "seeded beam." *Science* **187**, 642–644.

Ashford, J. R. and Sowden, R. R. [1970]. Multivariate probit analysis. *Biometrics* **26**, 535–546.

Bartlett, M. S. [1935]. Contingency table interactions. *J. Roy. Statist. Soc. Suppl.* **2**, 248–252.

Beaton, A. E. [1975]. The influence of education and ability on salary and attitudes. In F. T. Juster (ed.), *Education, Income, and Human Behavior*, pp. 365–396. New York, McGraw-Hill.

Benedetti, J. K. and Brown, M. B. [1976]. Alternative methods of building log-linear models. Unpublished manuscript.

Berkson, J. [1944]. Application of the logistic function to bio-assay. *J. Amer. Statist. Assoc.* **39**, 357–365.

Berkson, J. [1946]. Approximation of chi-square by "Probits" and by "Logits." *J. Amer. Statist. Assoc.* **41**, 70–74.

Bhapkar, V. P. and Koch, G. [1968]. On the hypotheses of "no interaction" in contingency tables. *Biometrics* **24**, 567–594.

Birch, M. W. [1963]. Maximum likelihood in three-way contingency tables. *J. Roy. Statist. Soc. B* **25**, 220–233.

Bishop, Y. M. M. [1967]. Multidimensional contingency tables: cell estimates. Ph.D. dissertation, Department of Statistics, Harvard University. Available from University Microfilm Service.

Bishop, Y. M. M. [1969]. Full contingency tables, logits, and split contingency tables. *Biometrics* **25**, 383–400.

Bishop, Y. M. M. [1971]. Effects of collapsing multidimensional contingency tables. *Biometrics* **27**, 545–562.

Bishop, Y. M. M. and Fienberg, S. E. [1969]. Incomplete two-dimensional contingency tables. *Biometrics* **25**, 119–128.

Bishop, Y. M. M., Fienberg, S. E., and Holland, P. W. [1975]. *Discrete Multivariate Analysis: Theory and Practice.* Cambridge, Mass., The MIT Press.

Bishop, Y. M. M. and Mosteller, F. [1969]. Smoothed contingency-table analysis. In J. P. Bunker et al. (eds.), *The National Halothane Study*, pp. 237–286. Report of the Subcommittee on the National Halothane Study of the Committee on Anesthesia, Division of Medical Sciences, National Academy of Sciences–National Research Council. National Institutes of Health, National Institute of General Medical Sciences, Bethesda, Md. Washington, D.C., U.S. Government Printing Office.

Blalock, H. M., Jr. [1964]. *Causal Inferences in Nonexperimental Research.* Chapel Hill, Univ. of North Carolina Press.

Blalock, H. M., Jr. [1972]. *Social Statistics* (2nd edition). New York, McGraw-Hill.

Bliss, C. I. [1967]. *Statistics in Biology*, Vol. 1. New York, McGraw-Hill.

Bloomfield, P. [1974]. Transformations for multivariate binary data. *Biometrics* **30**, 609–618.

Blyth, C. R. [1972]. On Simpson's paradox and the sure-thing principle. *J. Amer. Statist. Assoc.* **67**, 364–366.

Bock, R. D. [1970]. Estimating multinomial response relations. In R. C. Bose et al. (eds.), *Contributions to Statistics and Probability*, Chapel Hill, Univ. of North Carolina Press.

Bock, R. D. [1975]. Multivariate analysis of qualitative data. Chapter 8 of *Multivariate Statistical Methods in Behavioral Research*, New York, McGraw-Hill.

Brown, D. T. [1959]. A note on approximations to discrete probability distributions. *Information and Control* **2**, 386–392.

Brown, M. B. [1976]. Screening effects in multidimensional contingency tables. *Appl. Statist.* **25**, 37–46.

Brunswick, A. F. [1971]. Adolescent health, sex, and fertility. *Amer. J. Public Health* **61**, 711–720.

Bunker, J. P., Forrest, W. H., Jr., Mosteller, F., and Vandam, L. [1969]. *The National Halothane Study*. Report of the Subcommittee on the National Halothane Study of the Committee on Anesthesia, Division of Medical Sciences, National Academy of Sciences–National Research Council, National Institutes of Health, National Institute of General Medical Sciences, Bethesda, Md. Washington, D.C., U.S. Government Printing Office.

Chen, T. T. and Fienberg, S. E. [1974]. Two-dimensional contingency tables with both completely and partially cross-classified data. *Biometrics* **30**, 629–642.

Chen, T. T. and Fienberg, S. E. [1976]. The analysis of contingency tables with incompletely classified data. *Biometrics* **32**, 133–144.

Clayton, D. G. [1974]. Some odds ratio statistics for the analysis of ordered categorical data. *Biometrika* **61**, 525–531.

Cochran, W. [1954]. Some methods for strengthening the common $\chi^2$ tests. *Biometrics* **10**, 417–451.

Coleman, J. S. [1964]. *Introduction to Mathematical Sociology*. New York, Free Press.

Conover, W. J. [1974]. Some reasons for not using the Yates continuity correction on $2 \times 2$ contingency tables (with comments and a rejoinder). *J. Amer. Statist. Assoc.* **69**, 374–382.

Cornfield, J. [1956]. A statistical problem arising from retrospective studies. In J. Neyman (ed.), *Proceedings of the Third Berkeley Symposium on Mathematical Statistics and Probability*, Vol. 4, pp. 135–148. Berkeley, Univ. of California Press.

Cornfield, J. [1962]. Joint dependence of risk of coronary heart disease on serum cholesterol and systolic blood pressure: a discriminant function analysis. *Federation Proc.* **21**, 58–61.

Cox, D. R. [1966]. Some procedures connected with the logistic qualitative response curve. In F. N. David (ed.), *Research Papers in Statistics: Essays in Honour of J. Neyman's 70th Birthday*. pp. 55–71. New York, John Wiley.

Cox, D. R. [1970a]. *Analysis of Binary Data*. London, Methuen.

Cox, D. R. [1970b]. The continuity correction. *Biometrika* **57**, 217–219.

Cox, D. R. [1972]. The analysis of multivariate binary data. *Appl. Statist.* **21**, 113–120.

Cox, D. R. and Hinkley, D. V. [1974]. *Theoretical Statistics*. London, Chapman & Hall, and New York, Halsted (Wiley).

Cox, D. R. and Lauh, E. [1967]. A note on the graphical analysis of multidimensional contingency tables. *Technometrics* **9**, 481–488.

Daniel, C. [1959]. Use of half-normal plots in interpreting factorial two-level experiments. *Technometrics* **1**, 311–341.

Darroch, J. N. [1962]. Interaction in multi-factor contingency tables. *J. Roy. Statist. Soc. Ser. B* **24**, 251–263.

Darroch, J. N. [1974]. Multiplicative and additive interaction in contingency tables. *Biometrika* **61**, 207–214.

Darroch, J. N. [1976]. No-interaction in contingency tables. Unpublished manuscript.

Darroch, J. N. and Ratcliff, D. [1972]. Generalized iterative scaling for loglinear models. *Ann. Math. Statist.* **43**, 1470–1480.

Dawber, T. R., Kannel, W. B., and Lyell, L. P. [1963]. An approach to longitudinal studies in a community: the Framingham study. *Ann. N.Y. Acad, Sci.* **107**, 539–556.

Deming, W. E. and Stephan, F. F. [1940a]. On a least squares adjustment of a sampled frequency table when the expected marginal totals are known. *Ann. Math. Statist.* **11**, 427–444.

Deming, W. E. and Stephan, F. F. [1940b]. The sampling procedure of the 1940 population census. *J. Amer. Statist. Assoc.* **35**, 615–630.

Dempster, A. P. [1971]. An overview of multivariate analysis. *J. Multivariate Anal.* **1**, 316–346.

Dempster, A. P. [1973]. Aspects of the multinomial logit model. In P. R. Krishnaiah (ed.), *Multivariate Analysis: Proceedings of the Third International Symposium*, pp. 129–142. New York, Academic Press.

Draper, N. R. and Smith, H. [1966]. *Applied Regression Analysis*. New York, John Wiley.

DuMouchel, W. [1975]. On the analogy between linear and log-linear regression. Unpublished manuscript.

Duncan, O. D. [1966]. Path analysis: sociological examples. *Amer. J. Sociol.* **72**, 1–16.

Duncan, O. D. [1975a]. *Structural Equations Models.* New York, Academic Press.

Duncan, O. D. [1975b]. Partitioning polytomous variables in multiway contingency analysis. *Social Science Res.* **4**, 167–182.

Dyke, G. V. and Patterson, H. D. [1952]. Analysis of factorial arrangements when the data are proportions. *Biometrics* **8**, 1–12.

Dykes, M. H. M. and Meier, P. [1975]. Ascorbic acid and the common cold: evaluation of its efficacy and toxicity. *J. Amer. Med. Assoc.* **231**, 1073–1079.

Feller, W. [1968]. *An Introduction to Probability Theory and Its Application*, Vol. I (3rd edition). New York, John Wiley.

Fienberg, S. E. [1969]. Preliminary graphical analysis and quasi-independence for two-way contingency tables. *Appl. Statist.* **18**, 153–165.

Fienberg, S. E. [1970a]. An iterative procedure for estimation in contingency tables. *Ann. Math. Statist.* **41**, 907–917.

Fienberg, S. E. [1970b]. The analysis of multidimensional contingency tables. *Ecology* **51**, 419–433.

Fienberg, S. E. [1970c]. Quasi-independence and maximum likelihood estimation in incomplete contingency tables. *J. Amer. Statist. Assoc.* **65**, 1610–1616.

Fienberg, S. E. [1972a]. The analysis of incomplete multi-way contingency tables. *Biometrics* **28**, 177–202.

Fienberg, S. E. [1972b]. The multiple-recapture census for closed populations and incomplete $2^k$ contingency tables. *Biometrika* **59**, 591–603.

Fienberg, S. E. and Larntz, K. [1976]. Loglinear representation for paired and multiple comparisons models. *Biometrika* **63**, 245–254.

Fienberg, S. E. and Mason, W. M. [1977]. Identification and estimation of age–period–cohort effects in the analysis of discrete archival data. Unpublished manuscript.

Fisher, F. M. [1966]. *The Identification Problem in Econometrics.* New York, McGraw-Hill.

Fisher, R. A. [1950]. The significance of deviations from expectation in a Poisson series. *Biometrics* **6**, 17–24.

Fleiss, J. L. [1973]. *Statistical Methods for Rates and Proportions.* New York, John Wiley.

Gart, J. J. [1971]. The comparison of proportions: a review of significance tests, confidence intervals, and adjustments for stratification. *Rev. Int. Statist. Inst.* **39**, 148–169.

Gokhale, D. V. [1971]. An iterative procedure for analysing log-linear models. *Biometrics* **27**, 681–687.

Gokhale, D. V. [1972]. Analysis of loglinear models. *J. Roy. Statist. Soc. Ser. B* **34**, 371–376.

Goldberger, A. S. [1964]. *Econometric Theory*. New York, John Wiley.

Good, I. J. [1963]. Maximum entropy for hypotheses formulation especially for multidimensional contingency tables. *Ann. Math. Statist.* **34**, 911–934.

Good, I. J. [1975]. The number of hypotheses of independence for a random vector or for a multidimensional contingency table, and the Bell numbers. *Iranian J. Science and Technology* **4**, 77–83.

Goodman, L. A. [1963]. On methods for comparing contingency tables. *J. Roy. Statist. Soc. A* **126**, 94–108.

Goodman, L. A. [1964]. Simultaneous confidence limits for cross-product ratios in contingency tables. *J. Roy. Statist. Soc. Ser. B* **26**, 86–102.

Goodman, L. A. [1968]. The analysis of cross-classified data: independence, quasi-independence, and interaction in contingency tables with or without missing cells. *J. Amer. Statist. Assoc.* **63**, 1091–1131.

Goodman, L. A. [1969]. On partitioning $X^2$ and detecting partial association in three-way contingency tables. *J. Roy. Statist. Soc. Ser. B* **31**, 486–498.

Goodman, L. A. [1970]. The multivariate analysis of qualitative data: interactions among multiple classifications. *J. Amer. Statist. Assoc.* **65**, 226–256.

Goodman, L. A. [1971a]. The analysis of multidimensional contingency tables: stepwise procedures and direct estimation methods for building models for multiple classifications. *Technometrics* **13**, 33–61.

Goodman, L. A. [1971b]. Partitioning of chi-square, analysis of marginal contingency tables, and estimation of expected frequencies in multidimensional contingency tables. *J. Amer. Statist. Assoc.* **66**, 339–344.

Goodman, L. A. [1972]. A general model for the analysis of surveys. *Amer. J. Sociol.* **77**, 1035–1086.

Goodman, L. A. [1973a]. Causal analysis of data from panel studies and other kinds of surveys. *Amer. J. Sociol.* **78**, 1135–1191.

Goodman, L. A. [1973b]. The analysis of multidimensional contingency tables when some variables are posterior to others: a modified path analysis approach. *Biometrika* **60**, 179–192.

Goodman, L. A. [1975]. A new model for scaling response patterns: an application of the quasi-independence concept. *J. Amer. Statist. Assoc.* **70**, 755–768.

Goodman, L. A. and Kruskal, W. H. [1954]. Measures of association for cross-classifications. *J. Amer. Statist. Assoc.* **49**, 732–764.

Goodman, L. A. and Kruskal, W. H. [1959]. Measures of association for cross-classifications, II: further discussion and references. *J. Amer. Statist. Assoc.* **54**, 123–163.

Goodman, L. A. and Kruskal, W. H. [1963]. Measures of association for cross-classifications, III: approximate sampling theory. *J. Amer. Statist. Assoc.* **58**, 310–364.

Goodman, L. A. and Kruskal, W. H. [1972]. Measures of association for cross-classifications, IV: simplification of asymptotic variances. *J. Amer. Statist. Assoc.* **67**, 415–421.

Grizzle, J. E. [1961]. A new method of testing hypotheses and estimating parameters for the logistic model. *Biometrics* **17**, 372–385.

Grizzle, J. E. [1967]. Continuity correction in the $\chi^2$ test for $2 \times 2$ tables. *Amer. Statist.* **21**, 28–32.

Grizzle, J. E., Starmer, C. F., and Koch, G. G. [1969]. Analysis of categorical data by linear models. *Biometrics* **25**, 489–504.

Grizzle, J. E. and Williams, O. D. [1972]. Loglinear models and tests of independence for contingency tables. *Biometrics* **28**, 137–156.

Haberman, S. J. [1972]. Loglinear fit for contingency tables (Algorithm AS 51). *Appl. Statist.* **21**, 218–225.

Haberman, S. J. [1973a]. Loglinear models for frequency data: sufficient statistics and likelihood equations. *Ann. Math. Statist.* **1**, 617–632.

Haberman, S. J. [1973b]. Printing multidimensional tables (Algorithm AS 57). *Appl. Statist.* **22**, 118–126.

Haberman, S. J. [1974a]. *The Analysis of Frequency Data.* Chicago, Univ. of Chicago Press.

Haberman, S. J. [1974b]. Loglinear models for frequency tables with ordered classifications. *Biometrics* **30**, 589–600.

Hansen, W. L., Weisbrod, B. A., and Scanlon, W. J. [1970]. Schooling and earnings of low achievers. *Amer. Econ. Rev.* **50**, 409–418.

Hoaglin, D. [1977]. Exploring a table of percentage points of $\chi^2$. *J. Amer. Statist. Assoc.* **72** (in press).

Hocking, R. R. and Oxspring, H. H. [1974]. The analysis of partially categorized contingency data. *Biometrics* **30**, 469–484.

Ireland, C. T. and Kullback, S. [1968]. Contingency tables with given marginals. *Biometrika* **55**, 179–188.

Jennrich, R. I. and Moore, R. H. [1975]. Maximum likelihood estimation by means of nonlinear least squares. Tech. Report No. 3, Health Sciences Computing Facility, University of California, Los Angeles.

Kastenbaum, M. [1974]. Analysis of categorized data: some well-known analogues and some new concepts. *Comm. Statist.* **3**, 401–417.

Killion, R. A. and Zahn, D. A. [1976]. A bibliography of contingency table literature: 1900–1974. *Int. Statist. Rev.* **44**, 71–112.

Kimura, M. [1965]. Some recent advances in the theory of population genetics. *Japan. J. Hum. Genet.* **10**, 43–48.

Koch, G. G., Freeman, D. H., Jr., and Tolley, H. D. [1975]. The asymptotic covariance structure of estimated parameters from contingency table log-linear models. Unpublished manuscript.

Kruskal, W. H. [1968]. Tests of significance. *Int. Encyclopedia of the Social Sciences* **14**, 238–250.

Ku, H. H. and Kullback, S. [1968]. Interaction in multidimensional contingency tables: an information theoretic approach. *J. Res. Nat. Bur. Standards* **72B**, 159–199.

Ku, H. H. and Kullback, S. [1974]. Loglinear models in contingency table analysis. *Amer. Statist.* **28**, 115–122.

Ku, H. H., Varner, R. N., and Kullback, S. [1971]. Analysis of multidimensional contingency tables. *J. Amer. Statist. Assoc.* **66**, 55–64.

Kullback, S. [1973]. Estimating and testing linear hypotheses on parameters in the log-linear model. *Biomet. Zeit.* **15**, 371–388.

Kullback, S., Kupperman, M., and Ku, H. H. [1962]. An application of information theory to the analysis of contingency tables with a table of $2N \ln N$, $N = 1(1)\,10{,}000$. *J. Res. Nat. Bur. Standards* **66B**, 217–243.

Lancaster, H. O. [1951]. Complex contingency tables treated by the partition of chi-square. *J. Roy. Statist. Soc. Ser. B* **13**, 242–249.

Lancaster, H. O. [1957]. Some properties of the bivariate normal distribution considered in the form of a contingency table. *Biometrika* **49**, 289–292.

Lancaster, H. O. [1969]. *The Chi-squared Distribution*, chapters 11 and 12. New York, John Wiley.

Lancaster, H. O. and Hamdan, M. A. [1964]. Estimation of the correlation coefficient in contingency tables with possible nonmetrical characters. *Psychometrika* **29**, 381–391.

Larntz, K. [1973], Small sample comparisons of chi-square statistics. *Amer. Statist. Assoc. Proc., Social Statist. Sect.* 336–339.

Larntz, K. and Weisberg, S. [1976]. Multiplicative models for dyad formation. *J. Amer. Statist. Assoc.* **71**, 455–461.

Lee, S. K. [1975]. On asymptotic variances for loglinear models in contingency tables. Unpublished Ph.D. dissertation, School of Statistics, University of Minnesota.

Lindley, D. V. [1964]. The Bayesian analysis of contingency tables. *Ann. Math. Statist.* **35**, 1622–1643.

Lindley, D. V. and Novick, M. R. [1975]. Inferences from large data sets to individual cases: use of appropriate conditional probabilities in exchangeable subpopulations. Unpublished manuscript.

Lindsey, J. K. [1973]. *Inferences from Sociological Survey Data: A Unified Approach.* New York, Elsevier.

Lombard, H. L. and Doering, C. R. [1947]. Treatment of the four-fold table by partial correlation as it relates to public health problems. *Biometrics* **3**, 123–128.

Manski, C. F. and Lerman, S. R. [1976]. The estimation of choice probabilities from choice based samples. Unpublished manuscript.

Manski, C. F. and McFadden, D. [1977]. Alternative estimators and sample designs for discrete choice analysis. Unpublished manuscript.

Maxwell, A. E. [1961]. *Analysing Qualitative Data.* London, Methuen.

McFadden, D. [1976]. Quantal choice analysis: a survey. *Annals of Economic and Social Measurement* **5**, 363–390.

Meehl, P. E. and Rosen, A. [1955]. Antecedent probability and the efficiency of psychometric signs, patterns, or cutting scores. *Psychol. Bull.* **52**, 194–216.

Mosteller, F. [1951]. Remarks on the method of paired comparisons, III: a test for paired comparisons when equal standard deviations and equal correlations are assumed. *Psychometrika* **16**, 207–218.

Mosteller, F. [1968]. Association and estimation in contingency tables. *J. Amer. Statist. Assoc.* **63**, 1–28.

Nelder, J. A. [1974]. Log-linear models for contingency tables: a generalization of classical least squares. *Appl. Statist.* **23**, 323–329.

Nelder, J. A. [1976]. Hypothesis testing in linear models (Letter to the Editor). *Amer. Statist.* **30**, 101.

Nelder, J. A. and Wedderburn, R. W. M. [1972]. Generalized linear models. *J. Roy. Statist. Soc. Ser. A* **135**, 370–384.

Nerlove, M. and Press, S. J. [1973]. Univariate and multivariate log-linear and logistic models. Rand Corporation Tech. Report R–1306–EDA/NIH, Santa Monica, Calif.

Pauling, L. [1971]. The significance of the evidence about ascorbic acid and the common cold. *Proc. Natl. Acad. Sci.* (*USA*) **68**, 2678–2681.

Pearson, E. S. and Hartley, H. O. [1966]. *Biometrika Tables for Statisticians, Volume 1* (3rd edition). London, Cambridge Univ. Press.

Pearson, K. [1900a]. On a criterion that a given system of deviations from the probable in the case of a correlated system of variables is such that it can be reasonably supposed to have arisen from random sampling. *Philos. Mag.* **50**, No. 5, 157–175.

Pearson, K. [1900b]. Mathematical contributions to the theory of evolution in the inheritance of characters not capable of exact quantitative measurement, VIII. *Philos. Trans. Roy. Soc. Ser. A* **195**, 79–150.

Pearson, K. and Heron, D. [1913]. On theories of association. *Biometrika* **9**, 159–315.

Pielou, E. C. [1969]. *An Introduction to Mathematical Ecology.* New York, John Wiley.

Plackett, R. L. [1964]. The continuity correction in $2 \times 2$ tables. *Biometrika* **5**, 327–337.

Plackett, R. L. [1965]. A class of bivariate distributions. *J. Amer. Statist. Assoc.* **60**, 516–522.

Plackett, R. L. [1974]. *The Analysis of Categorical Data*. London, Griffin.

Ploog, D. W. [1967]. The behavior of squirrel monkeys (*Saimiri sciureus*) as revealed by sociometry, bioacoustics, and brain stimulation. In S. Altman (ed.), *Social Communication Among Primates*, pp. 149–184. Chicago, Univ. of Chicago Press.

Rao, C. R. [1973]. *Linear Statistical Inference and Its Applications* (2nd edition). New York, John Wiley.

Reiss, I. L., Banwart, A., and Foreman, H. [1975]. Premarital contraceptive usage: a study and some theoretical explorations. *J. Marriage and the Family* **37**, 619–630.

Ries, P. N. and Smith, H. [1963]. The use of chi-square for preference testing in multidimensional problems. *Chemical Engineering Progress* **59**, 39–43.

Roy, S. N. and Kastenbaum, M. A. [1956]. On the hypothesis of no "interaction" in a multi-way contingency table. *Ann. Math. Statist.* **27**, 749–757.

Roy, S. N. and Mitra, S. K. [1956]. An introduction to some nonparametric generalizations of analysis of variance and multivariate analysis. *Biometrika* **43**, 361–376.

Savage, I. R. [1973]. Incomplete contingency tables: condition for the existence of unique MLE. In P. Jagars and L. Råde (eds.), *Mathematics and Statistics. Essays in Honour of Harold Bergström*, pp. 87–90. Göteborg, Sweden, Chalmers Institute of Technology.

Schoener, T. W. [1968]. The *anolis* lizards of Bimini: resource partitioning in a complex fauna. *Ecology* **49**, 704–726.

Sewell, W. H. and Shah, V. P. [1968]. Social class, parental encouragement, and educational aspirations. *Amer. J. Sociol.* **73**, 559–572.

Simon, G. [1974]. Alternative analyses for the singly-ordered contingency table. *J. Amer. Statist. Assoc.* **69**, 971–976.

Simpson, E. H. [1951]. The interpretation of interaction in contingency tables. *J. Roy. Statist. Soc. Ser. B* **13**, 238–241.

Snedecor, G. W. and Cochran, W. G. [1967]. *Statistical Methods* (6th edition). Ames, Iowa, Iowa State Univ. Press.

Sokal, R. R. and Rohlf, F. J. [1969]. *Biometry*. San Francisco, W. H. Freeman.

Stouffer, S. A. and Toby, J. [1951]. Role conflict and personality. *Amer. J. Sociol.* **56**, 395–406.

Sundberg, R. [1975]. Some results about decomposable (or Markov-type) models for multidimensional contingency tables: distribution of marginals and partitioning of tests. *Scand. J. Statist.* **2**, 71–79.

Talbot, J. and Mason, W. M. [1975]. Analysis of the 4-way table: one-third of a nation data. Unpublished manuscript.

Taubman, P. and Wales, T. [1974]. *Higher Education and Earnings: College as an Investment and a Screening Device*. New York, McGraw-Hill.

Thorndike, R. L. and Hagen, E. P. [1959]. *Ten Thousand Careers*. New York, John Wiley.

Truett, J., Cornfield, J., and Kannel, W. [1967]. A multivariate analysis of the risk of coronary heart disease in Framingham. *J. Chron. Dis.* **20**, 511–524.

Wakeley, P. C. [1954]. Planting the southern pines, *U.S. Dept. Agr. Forest Serv. Agr. Monogr.* **18**, 1–233.

Walker, S. H. and Duncan, D. B. [1967]. Estimation of the probability of an event as a function of several independent variables. *Biometrika* **54**, 167–179.

Wermuth, N. [1976a]. Analogies between multiplicative models in contingency tables and covariance selection. *Biometrics* **32**, 95–108.

Wermuth, N. [1976b]. Model search among multiplicative models. *Biometrics* **32**, 253–264.

Williams, O. D. and Grizzle, J. E. [1972]. Analysis of contingency tables having ordered response categories. *J. Amer. Statist. Assoc.* **67**, 55–63.

Wing, J. K. [1962]. Institutionalism in mental hospitals. *Brit. J. Soc. Clin. Psychol.* **1**, 38–51.

Wold, H. and Jureen, L. [1953]. *Demand Analysis*. New York, John Wiley.

Wright, S. [1960]. Path coefficients and path regressions: alternatives in complementary concepts? *Biometrics* **16**, 423–445.

Yates, F. [1934]. Contingency tables involving small numbers and the $\chi^2$-test. *J. Roy. Statist. Soc. Suppl.* **1**, 217–235.

Yule, G. U. [1900]. On the association of attributes in statistics: with illustration from the material of the childhood society, &c. *Philos. Trans. Roy. Soc. Ser. A* **194**, 257–319.

Yule, G. U. [1903]. Notes on the theory of association of attributes in statistics. *Biometrika* **2**, 121–134.

# Author Index

# Subject Index